REINVENTING YOURSELF

A Control Theory Approach to

Becoming the Person You Want to Be

REINVENTING YOURSELF

A Control Theory Approach to

Becoming the Person You Want to Be

D. Barnes Boffey, Ed.D.

New View Publications
Chapel Hill

Second printing

Reinventing Yourself: A Control Theory Approach to Becoming the Person You Want to Be. Copyright © 1993, D. Barnes Boffey Ed.D.

ISBN 0-944337-14-7

Library of Congress Catalog Card Number: 92-50851

Collages by Eugenia Morrison

Cover design by Jeffrey Hale and Fred Good

Photo of author by Fraser Randolph

Quantity Purchases
Companies, professional groups, clubs, and other organizations may qualify for special terms when ordering quantities of this title. For information contact the Sales Department, New View Publications, P.O. Box 3021, Chapel Hill, N.C. 27515-3021.

Manufactured in the United States of America.

*This book is dedicated to
two of the finest young men I know,
my two sons, David and Adam.*

Acknowledgements

There are many who deserve acknowledgement for their contributions to this book. Primarily, I must acknowledge Dr. William Glasser whose ideas and teachings about Control Theory and Reality Therapy constitute the major psychological underpinnings for the reinventing process. Without his ideas about "total behavior," this book would never have been written.

There are also others who have helped in bringing these ideas to life. My heartfelt thanks go out to those below and the many others who have supported me in this effort.

To Bill Schreck and Perry Good for showing me the incredible power of unconditional love and support, especially in the creative process.

To Bill and Naomi Glasser for their personal support and to my colleagues in the world of Reality Therapy—Barbara Garner, Diane Gossen, Dick Pulk, Nancy Buck, Al Katz, Suzy Hallock, Shelley Brierley, and many others in Croatia, Slovenia, and Australia who have shared their caring, knowledge, and expertise.

To all those who share my friendship with Bill W. and their willingness to love me when I couldn't love myself.

To the counselors and campers of Camp Lanakila who, over the course of thirty-one years, have provided me with the finest role models I could imagine.

To Sharon Boffey, with whom I learned the power of taking responsibility for becoming the person I want to be in a long-term relationship.

To Heidi Dahlberg who urged me on even after reading a ponderous first draft and to Kelly Lojk and Nancy Boffey who urged me on as I drew closer to a finished product.

To Joani and Sofian who listened to my ranting and raving when the pieces of this book wouldn't come together.

To the staff at New View Publications and particularly to Eugenia Morrison who created the collages, to Jeffrey Hale and Fred Good who designed the cover, and to Kelly Lojk who edited this book.

Contents

"One doesn't discover new lands without consenting to lose sight of the shore for a very long time."

\- André Gide

Chapter 1

Repair, Remodel, or Reinvent

The owners of the lumberyard in my home town in Vermont recently began building a huge storage barn for their large machinery, tools, and building supplies. The construction proceeded quickly, and new walls and roof supports added to the impression that the building process was going well—the result would be a sturdy storage barn.

One day as I drove by, I saw the whole building had collapsed; the walls, supports, and frame were now lying on each other like a pile of fallen dominoes. Clearly, things were *not* going according to plan, although someone had obviously developed a blueprint which they assumed would produce a functional building.

This building process followed the same pattern many of us encounter as we try to build successful lives. At a very young age, we begin to develop a plan for our lives and a vision of what we would like to be. We draw from the resources available to us at the time—the ideas, thoughts, feelings, behaviors, patterns, and perspectives of our families. We begin to build a foundation and framework for our lives with the assumption that if we follow our blueprints everything will work out well.

Repair

Many people grow up in healthy and functional environments and develop firm foundations for their lives. They absorb from their parents and other significant adults the principles, values, and behaviors which successfully guide them through life. They are able to find the happiness they hope for when they act on these principles and grow up having a sense of self-control even in difficult situations. Problems generally arise because they have not followed their blueprints and minor adjustments are needed in order to handle these problems. Their basic foundations are strong, as are the lives they have built, but they need to make repairs along the way to adjust to the realities of life.

I recently had a client like this. His name was Carl, and he came to see me with his fiancé. They were having problems in clarifying and agreeing to some basic propositions for their upcoming marriage. They needed someone to help them create both-win alternatives for dealing with issues of freedom, money, expectations of each other as partners, and their future roles as step-parents.

They did eventually work out all the issues that came up, but what became clear during our discussion was that at a basic level Carl had all the values and attitudes he needed to make this whole process work. He simply needed to make some repairs in his behavior to work through the differences with his fiancé. He spoke well of his upbringing and the values he learned from his parents. He often related the happiness he experienced in life to the basic assumptions he had maintained since childhood. He was content with the choices he had made throughout his education, early adulthood, and even his first marriage. He had put his blueprints to work in the real world and found that they were satisfactory—he was basically a happy person. Carl had come to counseling to make some repairs, but he certainly had no need to question the basic foundation upon which he had built his life.

Remodel

For some people the basic truths and foundations upon which they live are quite stable, but they run into big problems as they encounter specific relationships and situations. These problems need more than minor repairs. Remodeling becomes necessary when a portion of your blueprints must be reworked. In this scenario, the majority of your "house" (your life) is fine, but one "room" is no longer functional. Remodeling your life means altering the design or structure of that room.

Susan, for example, had a healthy outlook and a strong and effective psychological foundation. She felt in control and content most of the time, liked who she was, and found that she was generally able to control her feelings, thoughts, and actions. She had grown up with parents that loved her, as well as each other, and the basic perspectives she had learned as a child had usually worked well as an adult. Susan had been married for a few years and was now having trouble working out some problems with her husband.

Susan shared with me her concerns about the increasing frequency, length, and intensity of arguments she and her husband were having. She felt a growing inability to tell him how she was really feeling because she did not want to upset him. She also had a gnawing sense of guilt when she made demands on her husband's time for child-rearing. She began to see this as a problem when she realized her sense of personal well-being and self-worth were eroding. Susan was in a situation where a few simple repairs would not do the trick. She needed to look deeper into the issues in her marriage, finding ways to bring her behavior in her marriage into congruence with the way she handled other areas of her life.

Susan's "house" was generally in order, but one "room" (her marriage) needed to be remodeled to be consistent with the values and assumptions in the rest of her life. She needed to learn some new behaviors so that she might regain self-control during the difficult moments in her relationship. She knew how to maintain self-respect in professional and social situations—only her behavior in her marriage was out of synch.

Overall, Susan had received good blueprints as a child, but her parents' marriage had been typical of relationships in the 1950s. She had internalized the idea of the husband as the breadwinner and decision-maker and the wife as "supporter of her man." Now, as an adult in the 1990s, she was faced with her own marriage and its more contemporary demands for equality, flexible professional and child-rearing roles, and mutual problem-resolution (rather than pronouncements—even gentle ones—by "the king of the castle"). With some new skills, she could work to adapt her role in her marriage and make it functional for her. Susan and her husband began marriage counseling and together developed a new model for their relationship. These plans mirrored what they both wanted rather than repeating the old marital patterns that she (and, as it turned out, he also) had absorbed as a child. She was able to remodel a portion of her life so that her behavior was consistent with the competent and independent person she was in the other areas of her life.

Reinvent

Repairing and remodeling resolve most of the problems in the lives of healthy, functional people. However, some of us find that our blueprints are faulty and we must go beyond remodeling to the more profound process of reinventing ourselves from the foundation up. Like the construction of the storage barn, we reach a stage in our development where our plans fall apart. We discover that the basic assumptions, principles, ideas, thoughts, and behaviors with which we are working will never lead to a

6

fully functional life. Our failure is not an inability to build according to plan or an unwillingness to work hard. The basic design and principles with which we are building are flawed.

In my own case, I bottomed out as an alcoholic at the age of twenty-nine. Without a doubt, the life I had envisioned had not materialized. In fact, I had become the opposite of everything I wanted to be. Instead of being honest, I was closed and deceitful. Instead of being a good father, I was manipulative and childish. Instead of being a caring and helpful friend, I was deserving of my friends' pity and contempt, and instead of being a good husband, I was acting like a selfish and moody adolescent. In spite of what I had originally thought were great blueprints, my life had deteriorated into a pile of rubble.

Tom was a client whose two marriages had fallen apart and he was feeling depressed and helpless. He had fared well in divorce court, but there was an emptiness he felt inside himself when he thought of all the dreams and plans that had "gone up in smoke." His continuous blaming of his former wives for how they had ruined his life was sounding more and more hollow, even to Tom. As he endlessly chanted that others were always to blame, even his closest friends grew less interested in his tale of woe. He began to realize that his complaining was "getting old," and that perhaps he needed to resolve his problem in some other way. Tom started to assess his own role in the collapse of his most recent marriage. When he examined it closely, he began to perceive that this last marriage was just one of a long series of destructive relationships. This was new information for him, and he was startled when he became aware of his pattern of behavior. As he reviewed the way he had handled himself, particularly in intimate relationships, he suspected that he would now have to make some major changes if he wanted to correct the problem. He would have to look honestly at his own role in what he had previously thought was everyone else's problem, and develop new ways of behaving. Tom knew he wanted to learn healthy new ideas, but he was scared. He knew he had to learn how to be different, but he didn't know what "different" was.

As we talked about Tom's pattern of behavior and how he had learned to deal with people, especially women, it became increasingly clear that he had absorbed several "truths" at an early age which were now making a successful marriage very difficult. These included the beliefs that "you shouldn't show others how you really feel because then they can hurt you," "women are caretakers and men are to be taken care of," "admitting you are wrong decreases prestige," and "asking for help means you are weak." With these faulty assumptions as the blueprint for his life, Tom was unable to build an emotionally satisfying adult relationship.

As we catalogued the assumptions Tom was carrying into his relationships, he began to consider the depth of change that might be necessary. Tom was feeling very vulnerable and "stuck"; he wanted to alleviate his depression and pain as quickly as possible. In the immediacy of his confusion, the prospect of reinventing himself appeared to be too long and difficult.

Tom wished he could get by with a few repairs and get back to a new intimate relationship as soon as possible. He hoped he could simply "fix" the situation by taking a stress-free trip to the Caribbean "to get away from it all for awhile."

Tom also expressed the hope that if the vacation didn't work, he might be able to remodel himself. He could develop new strategies for coping with his failed marriages, but pursue the same general direction he had always followed. He could reinterpret his version of his marriage problems and create a new excuse for their failures. (If he switched the focus of his problems from women to the institution of marriage, he would still absolve himself of any major responsibility.) Tom could proceed short term without any major change in awareness or knowledge. With some quick repairs and remodeling, he imagined he could begin looking for a new relationship with a woman who would finally be "the right one for him."

Eventually, Tom agreed that repairing and remodeling would miss the mark completely. If he was really going to build a functional foundation for long-term relationships, he had to be willing to commit to a deeper level. He had to be willing to face his fears and reinvent himself.

Reinventing goes beyond repairing and remodeling to actually building a new structure. When you reinvent yourself, you create new beliefs, values, assumptions, and principles, and develop a way of living which unifies your behavior around these new foundations.

Chapter 2

Choosing Transformation

Reinventing ourselves is a transformational process through which we envision a future where we are meeting our basic needs and controlling our own behavior—where we are in balance regardless of what others are doing. It is a process which demands that we change the primary beliefs and assumptions that underlie all of our thoughts and actions. This process is rarely initiated until we are completely frustrated with the structure of our lives.

The central focus of reinvention is not where we have been, but rather where we are and where we are going. In this chapter we will take a look at the process which leads to the realization that we need to make profound changes in our lives. The rest of this book will focus on how to make these changes and how to continue to build on new foundations.

Ineffective Behaviors: Short-Term Success and Long-Term Consequences

It should be of no surprise to anyone that the majority of the concepts, assumptions, principles, and beliefs we use in our adult lives come from our families in the early stages of our development. As children, when we first experiment with a new behavior, we are not able to see the long-term consequences of those choices. We choose this behavior because we believe that it will enable us to meet our needs, and it usually does—in the short run. For example, when a child first discovers that lying can sometimes get her out of trouble, it may seem to be a behavior with no negative consequences, so why not use it?

As this child develops a pattern of lying, she will experience difficulties with friends and family which had not occurred before. She may also begin to feel a growing sense of inner unrest and discomfort. It is not only the specific behavior that has impact in our lives, but the *overall direction* of using that behavior over time.

To use another example: Sandy and Frank were recently married and had made a commitment to be accepting and supportive in their relationship. They had vowed to share their thoughts and feelings and to deal with each other lovingly, without sarcasm or criticism. They had chosen a positive direction in their relationship.

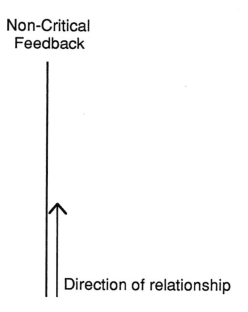

Non-Critical
Feedback

Direction of relationship

13

One day, Sandy asked Frank if he liked a new table-cloth she had bought, and he knew immediately that he didn't. He could tell from Sandy's voice that she really liked it, and he didn't want to hurt her feelings. He didn't want to lie, but it seemed like too much trouble to be frank, so he tried a new behavior. He made a slightly sarcastic remark, saying, "It's OK. It looks like something your mother would like." One might hear this remark and think, "What's the big deal?" This is an example of a slight change in the direction of non-critical feedback, say 10 degrees off the original course. Though this comment may not appear to be harmful to the relationship, a problem develops in ensuing incidents. When other sarcastic remarks get put in the place of non-critical feedback, a new direction is set. Even with a small 10 degree deviation, over time the distance from the original goal of supportive feedback increases dramatically.

Non-Critical
Feedback

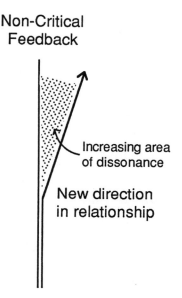

Increasing area
of dissonance

New direction
in relationship

After a few weeks, months, or years of the same kind
of remark, the difference between where they have
committed to be and where they are is quite large—large
enough to create stress and dissonance. If the pattern
continues, one day Sandy and Frank may find that communication between them has completely broken down.
The cumulative impact of many minor events is often
what triggers profound change.

The Search for an Easy Way Out

When we realize that our behaviors are not working, we may still believe we can find an easy answer that will not involve changing major patterns in our lives. We may attempt to do some minor repairs or remodel a small part of our behavior—anything to avoid going through the more difficult process of reinvention.

For example, compulsive overeaters often attempt to alter as many exterior factors as they can before they consider listening to their internal signals and changing how they meet their basic needs. They may try to choke off signals of malfunction and unmanageability with food. They may try every new diet that promises a quick fix to the problem, change what they eat, how many meals they eat, where they eat, and even how fast they eat in an effort to avoid looking at the central issues of their underlying beliefs and attitudes. When they have exhausted physical diets, they may listen to the school of thought which states that overweight people should just keep telling themselves that they are good, special people in spite of the way they look and eat. This is another form of denial. When people distort their bodies and use food as a drug, it is extremely unlikely that they will feel good about themselves.

Our intuition always leads us toward health, balance, and harmony. Working against this energy by abusing food leads to unsettling intuitive feelings. We can try to ignore our intuition, hoping that the problem will just go away, but the truth is that no amount of pleasure from food, sex, alcohol, or drugs will compensate for the unhappiness that we experience by ignoring our basic needs.

Like the compulsive overeater, many of us look to denial as an easy way out. We may attempt to avoid letting go of destructive assumptions. We may deny how badly we have felt at certain low points in our lives. We may deny that we consistently feel painful intuitive signals. Our denial works temporarily—it relieves our pain for moments or hours. But in the end, denial becomes as much a part of the problem as the original ineffective beliefs and behaviors.

Depression, like denial, may be another attempt to find an easy way out of our problems. Though we may *perceive* depression as a "virus we have caught," a disease over which we have very little control, it is actually a behavior that we choose. It may seem like an odd choice to make, but it often buys us time and shifts the focus off our own responsibility for fixing our problems. Depression, whether it involves sitting in a chair immobilized, sulking, complaining, being pessimistic, or contemplating suicide, may be our best attempt at the time to control the world. Although depression doesn't relieve our pain, it may win us sympathy from others and temporarily validate our claim of powerlessness. It may also be an attempt to preserve the belief that we are not responsible for how we feel and act.

A Series of Bottoms and a Moment of Truth

When we are ready to reinvent ourselves, it is generally because we have had some clear indication that the problems in our lives are rooted much more deeply than first suspected. In some cases, we must hit rock bottom before becoming aware of the need for reinvention.

As an active alcoholic, I was not willing to go beyond repairing or remodeling until I was separated from my family, in financial chaos, unemployable, out of contact with friends, an ineffective father, and fired from a summer job I had held for sixteen years. Only then was I willing to stop searching for an easy way out of my dilemma and look at the faulty blueprints I was using to make life choices.

It is hard to admit the truth when we believe that by doing so we may lose what we have and never be able to get what we want. However, we cannot begin a process of transformation until we come to a point where we admit that our behaviors or ideas are destructive to ourselves and others. We must accept that we are on the wrong path before we can look for a better one. How do we determine that our lives have finally become unmanageable? Is there some event in the external world that signals us that the "game is up"? Or is it a message that we receive internally?

Unhappy spouses, angry adolescents, stressed-out employees, and active alcoholics tell similar stories about what finally helped them face the unmanageability of their lives. They describe a wide variety of external events signifying their failure in the world—divorces, expulsions from school, hassles with bosses, lost jobs, and car accidents. But they also speak of an internal sense of hopelessness, stress, and pain which finally brought them to their knees. The individuality of each story makes it impossible to determine the absolute "last straw" which occurs before someone is willing to seek help. The significance of the event is relative to the meaning it has for the individual. To one overeater, finishing a whole half-gallon of ice cream may be "no big deal," to another overeater the same half-gallon may finally illustrate being out of control. One particular event, perhaps seemingly random, may finally tip the scale.

As a recovering alcoholic, many people have asked me why I finally stopped drinking. There was no singular incident that was "the reason" for my sobriety. Rather, there were many events which, like tiny weights on a balance scale, affected the equilibrium until I finally swung past the midpoint. Once, my four-year-old son, David, told a friend that instead of playing space wars or video games he wanted to go outside "to pretend to be drunk." I knew he had learned that from me. I also knew it was not what normal four-year-olds learn from their fathers. On several other occasions, I remember waking up in front of a blank television screen in the early morning after having passed out during a party. I have another memory of crying while watching a TV commercial which showed a drunk father embarrassing his children in front of their friends. These might appear to be small incidents in themselves, but each added its weight to the shifting balance. Eventually these incidents upset the equilibrium and what was manageable one minute became unmanageable the next.

My admissions of both powerlessness and unmanageability went hand in hand. As an active alcoholic, I continued drinking with the belief that I could choose how I wanted it to affect me, creating a myth about how powerful I was. I wanted to be able to drink heavily, treat myself and others in self-serving ways, and still feel good inside and loved by others. Reality doesn't work that way, but I didn't want to accept that. My admission of unmanageability came when I could no longer tolerate the internal pain created out of my external behaviors. It wasn't what happened to me; it was what the events showed me about myself which finally eroded my illusion of control. A major part of admitting the unmanageability of my life was becoming aware of certain physical, psychological, and spiritual truths which I had to take into account if I was to be happy and feel fulfilled.

Two Difficult Choices

It is not until we have exhausted all other avenues that we are willing to look at the steps involved in reinventing ourselves. If we still think there is an easy way out, or a way we can foist responsibility for our problems on to someone else, we are likely to avoid making the hard choices in front of us.

A friend of mine once told me, "Sometimes your only choices in life are to be knee deep in crap or ankle deep in crap." In all cases, we wish for more alternatives than we really have. Our minds repeatedly race through the facts and possibilities of the situation, hoping to find an easy answer that has been overlooked. *We see only two paths before us, but we desperately wish for a third.* Our first option is to continue on our current course—a path that we realize is becoming less and less effective. It is a well-marked path. The signpost ahead says: "The Same Path You've Been On: Certain Destruction."

The second path is marked with a sign that says "The Big Risk: Maybe Things Will Work Out and Maybe They Won't." It represents the only other real option to the way we have been proceeding. This path is one we have not traveled before and one which holds uncertainty. It is the unknown, but also our only real alternative to certain destruction.

We wish for a third path to appear, a path marked "Change Without Pain—Sure Success." This wish is part of a natural human desire to avoid suffering, a desire which may also lead us to use alcohol, drugs, food, or work as an escape. This third path is probably reformational and short-term in nature. Compulsive behaviors of any kind temporarily appear to open the "painless path," but in the long run the path doubles back on itself and joins the "Certain Destruction" path that we were trying to avoid initially.

Even as we begin to accept that these are our only two choices, we may still wish things were different—we wish for "more, better, yesterday." When we make difficult choices, there is almost always a period of time when we wish that we could turn back the clock to a time of ignorance and bliss, a time we remember warmly as "the way things used to be." But we must remember that if things had been as good as we now fantasize them to have been, we would have no need for change. Throughout the reinventing process, it will be necessary to look back with reason and ahead with faith. As we walk down the new path, we must remember we are not necessarily here because we want to be; we are here because the other alternatives are worse.

Creating New Foundations

Reinventing ourselves means taking responsibility for creating positive changes in our lives. We choose new foundational beliefs and learn how to unify our lives around these principles through effective behaviors. The way to change our lives is not by getting rid of what we don't want—but by creating what we do want. Eliminating negative behaviors doesn't necessarily create positive ones—it may just leave us in a neutral place. Profound change can only be sustained when we focus on what we *are doing*, not on what we *aren't doing*. We don't push away darkness, we turn on the light. We can waste a lot of energy by working to get rid of negatives. We may, for example, attempt to "stop feeling so guilty," but this will not bring the relief we desire. If we stop feeling guilty, we must start doing something else—it is impossible to *not* behave. Whether our new behavior is "being grateful" or "forgiving," learning to use it will eventually enable us to replace feeling guilty.

Old behaviors don't just "go away"; we stop using them because we choose better behaviors. When we walk into a room and turn on the light, where does the darkness go? The light pushes away the darkness because it is more powerful, but the darkness returns if the light is extinguished. If we should choose to stop using our new behaviors, our old behaviors—ineffective, yet always available to us—may once again appear to be our best (or only) options.

In twelve-step recovery programs the message is always that we are "recovering" overeaters or "recovering" alcoholics or "recovering" addicts. We are told that we will always retain our capacity to abuse our drug of choice, but by maintaining a positive and healthy style of living we can avoid returning to addictive behavior. As it says in the "Big Book" of Alcoholics Anonymous, "We are not cured of Alcoholism. What we have is a daily reprieve contingent on the maintenance of our spiritual condition." Addictive behavior is always an option for us, but as we practice other, more effective behaviors (being grateful, going to meetings, praying and meditating, keeping the faith, and being honest), we develop a new foundation for recovery. Reinventing ourselves is one way of creating the positive feelings, thoughts, and actions that are part of this new foundation.

Chapter 3

Reinventing Yourself

Reinventing yourself is a process in which you consciously decide which values and principles you want to live by. It is a process through which you go beyond any present confusion and dysfunction and visualize how your ideal self might feel, think, and act in the current situation.

In the reinventing process, you ask yourself, "If I had the courage to be the person I wanted to be, how would I handle my present situation?" The answer to that question then becomes the *unifying vision* of your behavior. After asking the question and visualizing the answer, the next step is to act as if you were the person you want to be rather than acting as you are now.

If you are in a situation where you feel scared and overwhelmed, the reinventing process would consist of the following stages. First, ask yourself, "If I were the person I wanted to be, how would I like to be feeling in this situation?" The answer is often difficult because you must first get beyond how you *do* feel in order to visualize how you might *like* to feel. Once you are able to envision your ideal self, you might see yourself as a person who is feeling calm, confident, and courageous.

The next step would be to ask yourself, "If I were calm, confident, and courageous, what would I be saying to myself? What would I be thinking?" Now develop the thoughts that your higher self would have. Again, imagining thoughts and feelings you aren't having can be very difficult (especially if you believe that outside forces control how you feel). By being creative and remembering previous positive experiences when you *did* have those feelings, you can usually create the thoughts that you believe your higher self would have in the current situation.

The next step in the process is to imagine the actions you would choose to take if you were feeling and thinking as you would like to be. In this stage, ask yourself the question, "If I were *feeling* calm, confident, and courageous and *thinking* 'I can do this,' and 'I have the courage to succeed here,' what actions would I then choose to take?" Visualize specific steps you could take if you were feeling as you wanted. The actions you then take will be unified by calm, confident, and courageous feelings.

The assumption here is that by focusing on the things you can directly control (thoughts and actions), you will eventually begin to feel the way you want. You will also be able to stay in control in those situations which you previously believed *caused* your feelings of fear or anxiety. Reinventing yourself is the process of unifying all thoughts and actions around the values that you want as your foundation. You act as if they were already in place, even though you may not "feel" like doing it. In twelve-step programs this is often defined as "Fake it 'til you make it." Some may see this as oversimplified, homespun psychology, but these ideas are strongly rooted in well-defined principles of human behavior.

Before going on to discuss these principles and their ramifications, it might be helpful to share a specific example of the reinventing process.

Marie was a client of mine who was very angry at her husband and felt no intimacy in their marriage. They constantly argued about an affair that her husband had ended three years earlier, and she felt her husband was extremely critical of her ideas and style of parenting.

Marie explained how she had tried everything "to make things better." She and her husband had been to a counselor two years earlier, she had moved out for a while, she had talked to her husband and told him to treat her with more consideration, and she had threatened him with a divorce if he didn't become a better partner. She related one incident in which she was so angry that she ended up beating on his chest and then breaking down and crying.

As she sensed her marriage disintegrating, she felt the rest of her life was also falling apart. Her kids wouldn't do what she asked and her friends didn't understand what she was going through. She was often depressed and angry. She looked very tired and said she just didn't know what to do anymore.

Marie had been working for years with the basic belief that she could "make" people, especially her husband, do what she wanted, and now she felt she had exhausted her alternatives. Marie had been feeling painful internal signals for a long time and had tried everything she could think of to repair and remodel her relationship. She admitted that no amount of screaming, crying, begging, or other coercive behaviors could "make him understand" or "make him stop criticizing me."

Marie was ready to reinvent herself and thereby change the foundation of her relationship. As we worked together, Marie evaluated her present behavior in her attempt to have a loving marriage. We drew a circle like the one on the next page and discussed the components of her total behavior: action, thinking, feeling, and physiology. We then went through a series of questions which focused on each of these four components and how they related to each other.

Present Behavior: Feelings

Marie described many of the hassles she was having with her husband in more detail. Once she had released some of her pent-up energy by sharing her frustration, I began to ask some simple questions. When asked how she was feeling in the situation with her husband, Marie admitted that she was feeling angry, depressed, and manipulated.

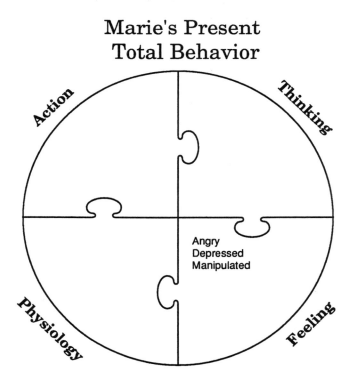

Marie's Present
Total Behavior

Thinking

After filling in her present feelings, I asked, "When you're feeling angry, depressed, and manipulated what are you thinking and saying to yourself? What is your self-talk?" Marie shared the following thoughts:

When I'm feeling angry, I'm thinking:
"My husband is an inconsiderate slob."
"Who the hell is he to judge me as a parent?"
"This isn't fair; I shouldn't be treated this way."
"It's not my fault; he's the one who should be in therapy."

When I'm feeling depressed, I'm thinking:
"There's nothing I can do that will get me out of this situation."
"Why bother? It's never going to get any better."
"I guess I'll never really have any love in my life."

When I'm feeling manipulated, I'm thinking:
"I never should have married Bill in the first place. I'm stupid."
"He's making me look bad in front of the kids."
"He knows I can't leave; he just stays around out of laziness."

Marie was in a situation that she could not handle effectively, and she was stuck in a cycle of anger, depression, and victimization that was now affecting all areas of her life. At this point we added her thoughts to the behavioral map.

Marie's Present Total Behavior

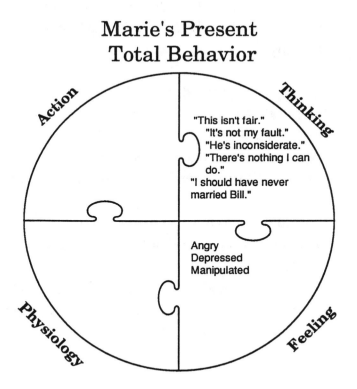

Actions

Next, Marie described the actions she had been taking to cope with and gain control of herself and her relationship with her husband. "What have you been doing when you feel angry, depressed, and manipulated and think these negative thoughts?" She described moping in her chair in the living room, complaining to her family and friends, arguing with her husband and kids about even the littlest issue, and frequently staying home rather than going out with friends. We then added Marie's actions to her behavioral chart.

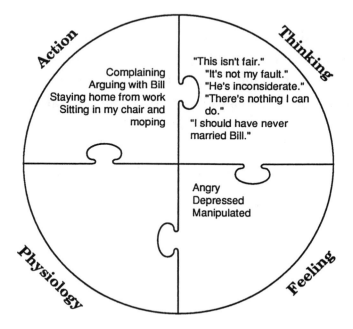

Physiology

As we continued to talk, Marie began to describe what happened in her body when she acted on these feelings and thoughts. What was her body telling her and what was her accompanying physiology in this situation? She described that she felt tense and had a knot in her stomach. In addition, her physician had told her she had high blood pressure.

Naming the Total Behavior

After adding her physiology to the behavioral map, I then asked, "How might you describe this whole package of feeling, thinking, action, and physiology? If you had to give all these pieces of behavior a name, what would you call it?" Marie answered, "Being Depressed."

Marie's Present
Total Behavior of
"Being Depressed"

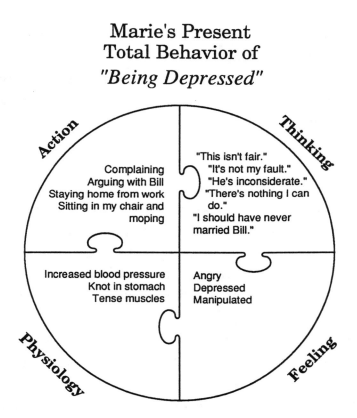

Marie could now look at her current behavior and decide two things. First, she could decide if what she was doing was working, and second, she could decide if this behavior was representative of the kind of person she wanted to be. Did she want her behavior to be unified by anger, depression, and victimization? Or would she like to be more in control of her own behavior? Looking at a behavioral map in a difficult situation may be an unpleasant experience because we can see clearly what we are doing and who we have become. Marie readily admitted two things about her behavior: what she was doing wasn't working to make things better, and even more importantly, she was disappointed with the person she had become.

Life constantly presents us with situations in which the world is not the way we want it to be. We feel frustrated in these situations, but it is nothing like the frustration and pain we feel when we perceive that we are no longer the people we want to be in those situations. We may have traded our standards, courage, and self-respect in an attempt to make things work out. We may have ignored our intuition and hoped that the ends would justify the means. When we perceive that we are not who we want to be, we begin to dislike ourselves. *When the world is not the way we want it to be—that can hurt us. When we are not the people we want to be—that can destroy us.*

Reinventing ourselves allows us to become the people we choose to be even in the most difficult situations. The ability to retain our self-respect is the real root of self-love, freedom, and personal power. We discover that we can feel powerful or loving or calm even in a situation in which we previously believed that other things or people were "making" us feel powerless, hateful, or outraged.

Marie had the courage to look at her life and admit that many years ago she had begun a pattern of choosing ineffective behaviors. She was acting on the assumption that other people were to blame for her unhappiness, and they were therefore responsible for making her happy. I asked Marie whether she imagined that a new acquaintance, a healthy and energetic person, would want to have a relationship with her in her present condition. "No," she said emphatically, "I can't even stand being around myself most of the time." I asked her if she would be willing to put some energy into becoming a person she liked before she spent any more time blaming others (especially her husband) for not living up to her expectations. She hemmed and hawed. "Are you comfortable having so much of your life focused on depression, anger, and the belief that your husband controls how you feel?" I asked, "Are you happy being the kind of person who deals with situations in the ways you are currently dealing with them, or would you like to have more power and clarity?" These were loaded questions, but they were also ones that Marie had asked herself intuitively many times before. At that moment, Marie took a major step towards true change in her life by looking in the mirror and stating that she didn't respect herself. She made another breakthrough by asking for help and admitting that she didn't know how to break out of the cycle.

Some might argue that the next step in Marie's progress would be to analyze where she had learned these patterns of behavior and what she would do to get rid of them. *However, the reinventing process is one which focuses on the solution, not the problem.* If you create and act on positive choices and effective behaviors, the negative patterns in your life will gradually diminish.

New Feelings

In Marie's case, it was time to visualize her higher self. In the process of reinventing, we used the same behavioral map, but we began with what Marie *wanted* to be feeling instead of what she *had been* feeling. I began by asking, "If you were the person you wanted to be, how would you be feeling *in this situation?*"

Marie replied, "If I were the person I wanted to be, I wouldn't be in this situation." The measure of being in effective control of our lives is not our skill at avoiding difficult situations, rather it is our ability to deal effectively with those situations and to respect our behavior regardless of external results. I then asked, "Assuming this difficult situation continues to exist, how would you like to be feeling as you deal with it?"

After some discussion, Marie decided that she would like to feel "not so wishy-washy," "not so upset all the time," and "less scared about the future." Although it took some gritting of teeth, she also admitted that if she were the person she wanted to be she would be "feeling more loving toward her husband." The gritting of teeth (action) was connected to her current resentment toward her husband (feeling) and the desire to make her love conditional on his changing how he acted—"I'll be more loving if he is more loving" (thought). Challenging her desire to make her behavior contingent on her husband's was an important step in evaluating her old attitudes, beliefs, principles, and assumptions.

The next step was to translate Marie's desired feelings into positive, rather than negative, terms. Except for "loving," all of her choices were about what she wanted "not to feel" rather than what she wanted to feel. We discussed that "don't wants" are not in the ideal world of our higher selves.

Identifying the positive side of what we want is not always easy; it takes a great deal of practice. We notice what is wrong more easily than what is right, and we can be seduced into focusing on getting rid of the bad things in our lives rather than creating the good. If we go to a building contractor and ask to have a house built for us, it is very ineffective to describe what we want the house to look like by pointing at other houses saying, "I don't want my house to look like that." The "do wants" must be articulated. In Marie's case, she ended up deciding that she wanted to feel loving, strong (her flip of "less wishy-washy"), calm (her flip of "less upset all the time"), and confident (her flip of "less scared about the future").

In moving toward the goals of wellness and fulfillment, Marie was now moving away from remodeling and toward reinvention. Remodeling might have included looking for other things that Marie could do to "make" her husband do what she wanted. A remodeling question might have been "What else can you do in this situation to make your husband be more considerate and intimate?" The question would have fallen within Marie's old foundational thinking, the belief that we can "make" others feel a certain way. For Marie, remodeling would have helped her look for new ways to succeed, but her success would have been measured by her husband's actions rather than by her ability to gain control over her own behaviors.

With a clear statement of how Marie wanted to be feeling, we were ready to continue the process.

Marie's Reinvented Total Behavior

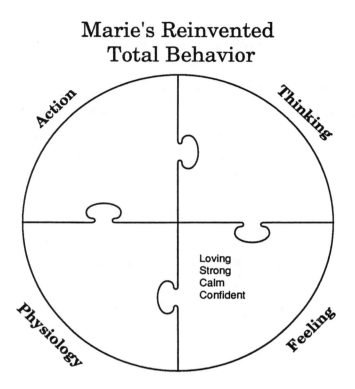

New Thoughts

"Marie, if you were feeling loving, strong, calm, and confident, what would you be thinking?" This question required Marie to describe the thinking behavior of her reinvented self. To find an answer, Marie needed to review the instances in her life where she had experienced those feelings and to recall the thoughts that accompanied them. If she could not remember these times in her own life, she could envision what a respected friend would be thinking.

Marie answered that she would be telling herself, "I can make a decision and stick to it," "I am an educated, competent woman who deserves respect," and "I can move forward in this situation even if it takes time." She began to create the thinking component of her reinvented behavior which was unified by the feelings of love, strength, calm, and confidence.

Marie developed more thoughts for each of her desired feelings and created the list below.

If I were feeling loving, I'd be thinking:
    "My husband is a good person who is acting
      badly."
    "I am a loving person who has a lot to give."
    "I need to take care of and be kind to myself."

If I were feeling strong, I'd be thinking:
    "This is a difficult time, but I can get through it."
    "I can make decisions and stick to them—I am
      strong."
    "I can be forgiving even when I feel hurt."

If I were feeling calm, I'd be thinking:
> "This too shall pass."
> "I am not going to get ruffled by others' behaviors."
> "I just need to walk through my fears."

If I were feeling confident, I'd be thinking:
> "This will work out."
> "God will help me if I need it."
> "I have people who believe in me and support me; I can go on."

These are not necessarily the only right thoughts for each of these feelings; they are Marie's thoughts. We talked about the consistency between her thoughts and feelings, and also whether these seemed congruent with what she perceived as her higher self.

After identifying Marie's reinvented thoughts, the thinking component of her reinvented behavior could be filled in on the chart.

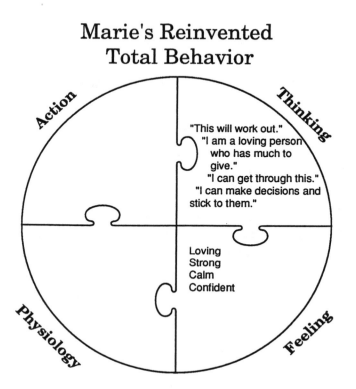

Marie's Reinvented Total Behavior

"This will work out."
"I am a loving person who has much to give."
"I can get through this."
"I can make decisions and stick to them."

Loving
Strong
Calm
Confident

Action

Thinking

Physiology

Feeling

New Actions

The next step in helping Marie reinvent herself is based on the question, "If you were feeling loving, strong, calm, and confident, and thinking the thoughts you have described, what would you be *doing*?" Marie used this question to create actions that were aligned with her new feelings and thoughts.

As Marie described these actions, she created a way to take some control of her life—regardless of what her husband chose to do. Marie said that if she were the person she wanted to be, she would tell her husband that she loved him and that she wanted things to work out, ask him to come to therapy, and talk with a lawyer about her legal rights. She said she would also get involved with swimming at the local pool rather than coming home and sitting in her chair all afternoon. When Marie saw that she had choices in the relationship, she felt a sense of freedom and a desire to take the next step. Now she could make specific plans about when she would ask her husband to begin marriage counseling, which lawyer she would see, and when she would go to the local pool. Marie had created a new blueprint for her own behavior which she could act on and in doing so, reinvent herself.

We then added the actions Marie would take as her reinvented self to the behavioral chart.

Marie's Reinvented
Total Behavior

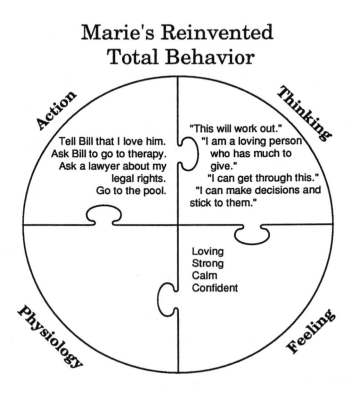

Action

Tell Bill that I love him.
Ask Bill to go to therapy.
Ask a lawyer about my
legal rights.
Go to the pool.

Thinking

"This will work out."
"I am a loving person
who has much to
give."
"I can get through this."
"I can make decisions and
stick to them."

Loving
Strong
Calm
Confident

Physiology

Feeling

New Physiology

In a situation in which we were primarily focused on changing our physiological behavior rather than our feelings, it would be very important to start with questions about physiology. We would start by asking, "What physical sensations would you like to experience?" and

"What would you like your body to be doing?" In a situation like Marie's, the answer might be "feeling less tense" (flip to "calm"), "not having headaches" (flip to "relaxed"), and "having normal blood pressure." We would then ask, "If you were relaxed and calm in your body, what would you be feeling?" and move on from there. In Marie's case, her primary focus was on what she was feeling emotionally, so that is the component of behavior which we addressed first. The more components of a reinvented behavior we can create and understand, the greater the chance of our integrating this behavior into our lives.

Naming Our Total Behavior

I continued: "So, Marie, it seems that you have the power to become the person you want to be even if your husband doesn't change a thing. That's a big step for someone who has been saying she can't stand being around herself most of the time. If you had to give this whole package of reinvented behavior a name, what would you call it?

Marie's answer was instantaneous: "Finally Respecting Myself." Her reinvented total behavior had a name, and Marie had a clear direction based on new unifying principles.

Marie's Reinvented Total Behavior of
"Respecting Myself"

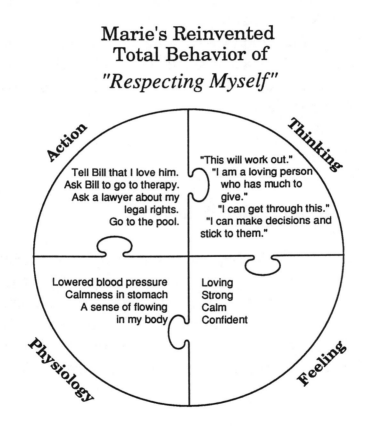

Action

Thinking

Tell Bill that I love him.
Ask Bill to go to therapy.
Ask a lawyer about my
legal rights.
Go to the pool.

"This will work out."
"I am a loving person
who has much to
give."
"I can get through this."
"I can make decisions and
stick to them."

Lowered blood pressure
Calmness in stomach
A sense of flowing
in my body

Loving
Strong
Calm
Confident

Physiology

Feeling

Just because we know what we would do if we were the people we wanted to be doesn't mean we can do all these things at once. We need to break down our overall direction into small, successful steps. *Any step forward* that is based on the unifying principles of love, strength, calmness, and confidence will be a positive and transformative behavior for Marie. Every time she acts on these principles, she increases these qualities in her life. As Mohandas Ghandi said, "Everything we do is insignificant, but it is very important that we do it."

Knowing the direction in which we need to move and actually taking steps in that direction are two separate things. In Marie's case, she had created a vision of how she would act if she were the person she wanted to be. This is the beginning of reinventing yourself, but the most difficult part of the process is actually *taking the action* and *continuing in that direction*. To live according to the principles and beliefs of a higher self is a tremendous on-going challenge.

Chapter 4

Getting What You Need

I have shared the basic concepts of the reinventing process with thousands of clients and workshop participants over the past years, and there have been a wide variety of comments and questions. Some have been happily overwhelmed with the simplicity of the concepts, while others have said, "If this is really as simple as you make it seem, why haven't people been taught to do this in counseling before? If it's as easy as you make it sound, why isn't this just the accepted way of changing our lives?"

In answer to these questions, the process is neither simple nor easy. It is hard work and takes a lot of energy over an extended period of time. It is exciting that in the early stages we can see a clear path where before there seemed to be only unending confusion. Communicating with one's higher self and creating concrete steps from that communication is very empowering. For someone who feels powerless, seeing a way to make things better is a welcome experience.

A major stumbling block in working with and believing in the concepts of reinvention is that they are based on a set of beliefs about human behavior which do not conform to "traditional wisdom." Conventional stimulus-response psychology assumes that someone or some set of circumstances outside of ourselves is causing our problems, and therefore we work to improve our ability to influence these people or circumstances. The focus is on people, places, or things that *make* us feel a certain way. For example, "*You* make me angry when you don't hold up your end of the bargain," or, "I'm depressed because *no one* here cares about me," or "I'm sad because *it's* the anniversary of my divorce." In all three cases, the responsibility for feeling as we do is tied to something or someone outside of ourselves.

The reinventing process is based on the primary assumption that we have control over all parts of our total behavior (including our feelings, thoughts, actions, and even much of our physiology). In a world where reinventing ourselves is possible, our problems always exist *within* us—as do the solutions. Our problems are the result of our perceiving, acting, thinking, and feeling. Changing these elements within ourselves allows us to be part of our own solutions. *It is important to understand that the entire focus of reinventing ourselves is on changing how we behave within a given situation, not changing the situation.*

The idea that we are the solution to our problems is an offshoot of Dr. William Glasser's concepts of Control Theory, a relatively new psychological model of human behavior.[1] I believe this theory describes the human condition more accurately than traditional stimulus-response

---

[1] Control Theory was initially developed as an engineering theory, but was later applied as a psychological model by William Powers in his book *Behavior: The Control of Perception.* Powers explains why living organisms act as they do. He developed the idea that organisms act because of internal motivating factors rather than because other people, places, or things "make" them do what they do. These ideas were then more fully developed by Dr. William Glasser in his books *Stations of the Mind* and *Control Theory*, by E. Perry Good in *In Pursuit of Happiness,* and by Ed Ford in *Freedom From Stress.*

psychology and that understanding Control Theory allows us to take giant steps toward being happy by creating more functional thoughts and actions.

Control Theory is a compass, rather than a map. It helps us understand human behavior in all situations, rather than defining exact behaviors for specific instances. A map is useful in clearly defined terrain. In uncharted territory we need a compass to maintain our true course. Control Theory is just that.

Before I learned Control Theory, I spent most of my time trying to get others to do what I wanted. I assumed they had to stop doing what was upsetting me before I could stop being upset. After learning Control Theory, I was able to take responsibility for both the problem and the solution by telling myself that ultimately *my* problem is *my* inability to find a way to meet *my* needs in the current situation. When I was teaching college, I had a colleague who was always in a sour mood. She was so critical and negative that I told people that she made me depressed, too. I spent a lot of time trying to cheer her up because I believed that I needed her to be more positive before I could be happy. After learning Control Theory, I now understand that my real problem was that I did not know how to be cheerful in a situation where someone else was so negative. Were I to begin the process of reinventing myself, I would assume that I could change how I felt whether she was happy or not. I would begin by asking myself, "If I were the person I wanted to be, how would I like to be feeling when I encounter such a negative and depressed person?" My answer would be "content, detached, and positive," and from there I could follow the process, inventing the thoughts and actions that would accompany those feelings. If I were able to do this, my colleague's temperament would no longer be a problem in my life.

Control Theory starts from a unique perspective about the basic "givens" in human behavior. As Dr. Glasser states:

> Control Theory explains why, and to a great extent how, all living organisms behave and that all we do all of our lives is behave. It contends that all of our behavior is purposeful and that purpose is always to attempt to satisfy basic needs that are built into our genetic structure. It is called Control Theory because all behavior is our best attempt to control ourselves (so that we can control the world we live in) as we continually try to satisfy one or more of these basic needs.

Control Theory begins with a simple premise—we all have both physical and psychological needs built into our genetic makeup. Our primary goal as human beings is to attempt to meet these needs as we encounter the people and situations in our lives.

It is universally accepted that we have physical needs for sleep, water, and food and that we must fulfill these needs if we are to survive and be healthy. If we go too long without meeting them, our bodies stop working effectively and there are many things that can and do go wrong.

Control Theory asserts that the same is true of our psychological needs, and it identifies these needs as:

- love and belonging
- power and influence
- fun and pleasure
- freedom and autonomy

These needs must be consistently met if we are to maintain good psychological health. We can ignore them for short periods of time, but in the long run we must attend to them. Like our physiological needs, when our psychological needs are not met, many things can go wrong. If we are to be balanced and fulfilled, we must meet these needs in a manner that does not stop others from meeting their own needs. *How* we meet our needs

57

will differ from person to person and culture to culture. *What* the needs are and the fact that we must meet them does not change.

In future discussion of these needs I will refer to them as spiritual as well as psychological. The "spiritual" aspect does not connote a religious context, but affirms the idea that these needs are part of our makeup as human beings. As Virginia Satir said, "What is obvious to me is that we did not create ourselves. . . Life is something inside of you. You did not create it. Once you understand that, you are in a spiritual realm."

Control Theory states that these four basic psychological/spiritual needs are universal. They are a set of instructions built into the structure of all human beings. Survival—meeting our physical needs—is a powerful force in our lives, but the need to stay alive is not our only motivation. The four psychological/spiritual needs also drive our behavior.

Love and Belonging

The need for love and belonging is perhaps the easiest for most people to accept as a universal psychological/spiritual need. We must be able to give and receive love and interact with people with whom we belong and feel secure. Within each of us is a psychological/spiritual instruction to connect with other humans—to touch each other so that we might feel fulfilled. Dr. Glasser says that we must have at least one other person in the world who loves us and who we love if we are to meet this need. When we love ourselves and other people, we feel "full" or "whole." To use a physiological analogy, when we are hungry it is because our bodies tell us we need food. When we eat well, we feel full and nourished. When we eat food that is not healthy, we don't meet our need for nourishment even though we have full stomachs. Psychologically, when we feel lonely, it is because we need love and a sense of belonging. When we find healthy ways to meet that need for love (family, friends, pets, organizations, communion with God), we feel full and nourished. When we attempt to meet that need in ways that are not healthy (abusive relationships, controlling behaviors, drugs and alcohol, sexual activity without intimacy), we will continue to feel lonely even though we think we should feel full. The need for love and belonging must be met in ways that are mutually fulfilling for all involved or the need will not be truly met.

Power and Influence

Power and influence is another of our basic needs. We each have an internal need for a sense of power—having an impact on the world in a way which allows us to act, accomplish, and create. If we walk into a room and switch on the lights, there needs to be sufficient *power* for the bulb to glow or we remain in darkness. In the world of people and social interaction, we need to have enough influence to be listened to and taken seriously. If we lack power in our lives, our system creates an intuitive signal (like hunger or loneliness) telling us that we are not fulfilling that need. We may receive this signal at various times: when we had hoped to perform a physical task but find that we are not strong enough, when we want to motivate our children to do well in school but they don't do their homework, or when we want others to think well of us in a work situation but they see us as incompetent. If we have little impact or are unable to accomplish what we hope, we experience powerlessness. There are many forms of power (competence, skill, prestige, "coolness," money, socioeconomic class, rank, wisdom), and there are many ways of attaining them. Like love, your need for power must be met in a way that does not stop others from meeting their own needs. Bosses who do not let employees make any decisions have a lot of power, but their employees feel unimportant. In

Control Theory organizations, effective bosses will find ways to meet their own needs and at the same time allow their employees to voice opinions, make decisions, and experience their own sense of accomplishment.

Fun and Pleasure

Another basic need is fun and pleasure. When we are not having fun and finding pleasure in our lives, we experience intuitive signals of boredom or sluggishness. To be balanced and fulfilled, we must experience some type of fun. This can be anything from specific activities (skiing, playing checkers, or dancing) to a more general sense of playfulness that is woven into all activities. Maintaining a sense of humor in difficult situations is also a sign of the ability to have fun. The need for fun, like power and love, must be met in healthy ways that do not impinge on others meeting their needs. For example, sarcastic humor may be funny to the perpetrator, but it will rarely create a sense of fun in the relationship if the other person feels attacked or criticized.

Freedom and Autonomy

The final psychological need identified by Control Theory is freedom—a sense of choice and autonomy. We must have options and be able to make decisions which allow us to meet our other needs effectively. We need privacy and the knowledge that we can choose to be who we want to be. If we don't fulfill this need, we experience a painful intuitive signal—the sensation of being trapped. This signal will not go away until we experience freedom in our lives, enough to help us gain effective control. We can have freedom "from" a situation (being far away from a sister we can't stand) or we can have freedom "in" a situation (feeling that we can be who we want even if our sister continues to be irritating).

Reinventing yourself begins by visualizing yourself meeting your needs in whatever situations you encounter. Often this is hard to do because you may feel off-balance and out of control. When I ask, "If you were the person you want to be, how would you be feeling in this situation?" I am really saying "Visualize yourself meeting your basic psychological/spiritual needs and tell me what you would be feeling, thinking, and doing." People tend to think of the situation as the problem. However, the true conflict is an inability to meet our needs. *When we are able to learn to meet our needs in a problem situation, that situation will no longer be a problem for us.*

Taking responsibility for your problems is part of what makes reinvention so difficult. It is hard to tell yourself that you are the solution to your problems. It is also difficult to deal with other people who have no intention of taking that kind of responsibility for themselves. However, when you choose to see yourself as the solution in any situation, you gain the self-control to do something about your behavior and get back into balance.

Chapter 5

The Intuitive "YES!"

Happiness, serenity, self-actualization, contentment, inner harmony—these are all words which describe an internal sense that we are safe in the world and able to maintain a need-fulfilling balance. We are all created with the ability to experience happiness.

Part of being human also means that we are each equipped with internal signals—intuition—which tell us whether or not we are meeting our basic psychological/spiritual needs. Intuition can tell you when you are in concert with your internal instructions or when you are entering a situation in which it will be difficult to meet your needs. These signals give a constant "readout" on your psychological/spiritual state and can be either painful or pleasurable. Shakti Gawain writes in her book *Living in the Light,* "The knowingness that resides in each of us can be accessed through what we call our intuition. By learning to contact, listen to, and act on our intuition, we can directly connect to the higher power of the universe and allow it to become our guiding force."

The greatest happiness we can experience occurs at those moments when three conditions coincide:

- The expectations we have for our own behaviors are being met or exceeded.

- Our internal needs for love, power, fun, freedom, and survival are being met.

- The actions of people and the realities of places and things in the outside world are matching what we want to be happening.

If we are the people that we want to be (fulfilling our personal expectations) and that the universe asks us to be (meeting our needs) at the same time that we are surrounded by people, places, and things which are pleasing to us (the outside world), there is a tremendous sense of unity. In this unity we feel happiness. However, the crucial alignment—the real foundation of happiness— is between our *internal instructions and our own behavior*. Our intuition informs us of this primary alignment. Our ability to create this alignment and therefore experience a positive internal signal—an intuitive "YES!"—is essential to a joyful life.

The intensity of an intuitive "YES!" increases with congruence among the three components and decreases with incongruence. Reinventing yourself is a process through which you learn to create the intuitive "YES!" by adjusting your actions so that they match your highest expectations and meet your needs. Very often, the third component, external reality, is the one over which you have the least control. It is also the one you need to be the least concerned with in an intuition-directed life. Your ability to be happy rests primarily on how well you deal with the world as you encounter it, not with the world itself. When the world is not as you want it to be, it is even more important to take actions which connect your needs for being loving, powerful, playful, and free with your highest personal expectations.

The following personal example of the reinventing process illustrates the important interaction of fulfilling expectations, meeting needs, and finding satisfaction in external reality.

Several years ago, my ex-wife and I were in the middle of our divorce. We were about to have a very heated and difficult session with our lawyers to begin discussing the realities of money, property, and children. I knew it would be very hard for me to deal with and for days ahead of time I felt scared, angry, depressed, and tense. I finally began to use the reinvention tools and asked myself, "If I were the person I want to be, what would I be feeling in this situation?" The qualities that came to me were: loving, honest, creative, and faithful (which I define as the opposite of fearful). I took the process one step further by asking, "If I were loving, honest, creative, and faithful, what would I be thinking and doing at that meeting with the lawyers?"

The answers didn't come right away and I struggled with bits and pieces of this question over the next few days. Finally I began to develop a picture of myself in that situation and how I would act if I were the person I wanted to be. I needed to reinvent myself if I was to finish that meeting and feel the way I wanted. I knew that I couldn't change how others would act in this situation. However, I could choose how *I* wanted to act. I wanted to act in a way that would bring me closer to meeting my needs and being the "person I wanted to be." I was not interested in repeating the patterns of behavior I had learned in childhood—choosing fear, my old standby behavior when I felt threatened or hurt.

I decided that I would be thinking: "This doesn't all have to happen in one day," "I don't have to make any decisions I'm not clear about," "I want to be loving toward my wife even though this is a difficult time," "There are a lot of ways we can settle this if we just take our time," and "If I am honest about how I feel without attacking others, things will eventually work out."

With these thoughts I developed a plan of action which included speaking calmly, taking a breath when I felt a jolt in my stomach, listening to what others were saying before answering, using the words "I disagree" rather than "You're wrong," and delaying decisions in areas where there was no agreement.

In the actual session, I was very successful at handling myself the way I had chosen, and I felt good about my behavior. I wasn't perfect and I didn't always know what to do, but when we had finished, I felt proud of myself for what I had done in one of the most difficult situations in my life. I experienced an exhilarating intuitive "YES!" As I became more confident that I could be loving, honest, creative, and faithful in my life, I was able to be more consistent in my behavior. Because I was more consistent in my behavior, I added an air of calm to the whole process. I now know that my calmness is one of the reasons that my former wife and I continue to respect and care about each other.

Total Behavior of
Being Myself

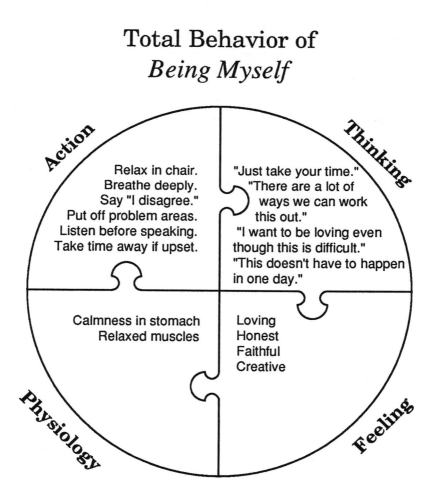

Action

Relax in chair.
Breathe deeply.
Say "I disagree."
Put off problem areas.
Listen before speaking.
Take time away if upset.

Thinking

"Just take your time."
"There are a lot of
ways we can work
this out."
"I want to be loving even
though this is difficult."
"This doesn't have to happen
in one day."

Physiology

Calmness in stomach
Relaxed muscles

Feeling

Loving
Honest
Faithful
Creative

There are difficulties in reinventing yourself, but in the long run the growth and joy are worth it. If you develop a way to live your life more closely tied to the guiding principles of your higher self, even in the most difficult of situations, you will find you have a sense of peace and self-esteem which you can carry wherever you go. As you develop this feeling of serenity by controlling yourself more fully, you are also more likely to influence the third component which increases your pleasure, your external world.

Chapter 6

The Transformational Shift

New Assumptions

Reinventing yourself is a process which takes vision, time, and hard work. In fact, after presenting these ideas at a workshop, one participant told me that I shouldn't talk about how difficult the process is because it could put people off. I understood what she was saying, but I would rather tell people the truth up front than underrate the difficulty of the process. Too often in psychological and self-help literature, there is an implied message that the process is easy if you only do it "right." In reality, if you choose to follow the path of reinventing yourself, there will be moments of doubt, confusion, fear, and indecision. However, by sticking with the process, the positive changes will be profound.

Acting on the assumption that the source of happiness is internal rather than external is new for many people. If you understand that you are making a major shift in your perception of reality and accept that there will be many unknowns, you can choose to act on faith, confidence, and hope instead of fear, doubt, and despair.

While repairing and remodeling are *reformational*, the reinventing process is *transformational*. You start by changing the foundations of your thinking and work up towards action. Marilyn Ferguson discusses this kind of change in her book *The Aquarian Conspiracy*: "Transformation . . . is the fourth dimension of change, the new perspective, the insight that allows the information to come together in a new form or structure." Whenever we go through a shift in our basic conceptual framework, the external outcomes are less predictable and therefore the process is more difficult. In reinventing yourself, you are acting on several new conceptual bases, therefore creating an entirely new structure.

First, you are taking responsibility for your problems by accepting that *you* are the solution. You accept that your psychological/spiritual instructions are to find a way to maintain your balance in the situations you encounter, not necessarily to change those situations.

Second, you are working with the principle that your behavior is made up of four components (action, thinking, feeling, and physiology), and that these components affect each other. Rather than assuming that feelings are caused by an external "stimulus," you are assuming that you feel the way you do because of the way you think and act. You are also accepting the principle that you have direct control over your thoughts and actions, and thus indirect control over your feelings and physiology.

Third, you are accepting that there are internal signals within you which give constant feedback on the state of your psychological/spiritual well-being. This system gives pleasurable signals when your needs are being met and sends painful ones when they are not. Your intuition is the feedback around which you unify your behavior. When you truly follow your intuition, you act more loving, powerful, playful, and free.

Fourth, you are assuming that your needs are best met when you behave in concert with your higher self—your vision of the person you want to be. You assume that if you live in congruence with your higher self you will be happy and in control of your life.

It is impossible to live pleasurably without living wisely, well, and justly, and impossible to live wisely, well, and justly without living pleasurably.
-Epicurus

The following pages describe the ramifications of following through on reinventing yourself. I have provided suggestions to help you understand the intuition-directed journey and make the process more successful.

Vision Into Action—Facing Your Fear

Once you have chosen to reinvent yourself and have defined the person you want to be, the next stage in the process is putting your reinvented blueprints into action. You start *acting on your vision* and facing the fear. You begin to take the steps that have been defined in the reinventing process, understanding that there are many times when you must take actions that feel very "phony" because you don't really feel like taking them. Through an understanding of how your behavior works you know that you must "act your way into a new way of feeling rather than feel your way into a new way of acting."

Imagine a scenario in which you are angry at a person whom you perceive is "making" you feel intimidated and rejected. You are afraid to face this person honestly because this person may put you down or dismiss your concerns. Finally, imagine that you come to the point where this situation is no longer acceptable to you and you decide to reinvent yourself. Rather than centering your life around the belief that other people can intimidate and control you, you now choose to focus on your own internal balance.

The values you choose will unify your behavior. If you decide to reinvent yourself around the principles of confidence, courage, and honesty, you would ask yourself, "If I were feeling these things, what would I be thinking? What would I be saying to myself?"

You might give some of the following answers:

If I were feeling confident, I would be saying to myself:
"I can handle this situation."
"I have enough knowledge to make this work."
"I can do whatever I put my mind to."

If I were feeling courageous, I would be saying to myself:
"I will not be destroyed by fear."
"Yes, this is scary, but other people have done it and so will I."
"I will stay focused on my progress and continue in a positive direction."

If I were feeling honest, I would be saying to myself:
"This is scary."
"Here's who I am; you can accept me or not, but this is me."
"This is how I feel and this is what I think, and I'm going to tell myself and others the truth."
"I need to make sure I am ready to take care of myself regardless of whether this person changes his behavior."

With these thoughts in mind, ask yourself, "If I were the person I want to be, feeling confident, courageous, and honest and thinking the thoughts above, what would I do?"

When acting on your vision, you may feel scared, unsure, and self-critical. Act on the knowledge that the feelings you desire will eventually come, and practice new behaviors even though they may feel unnatural. Acting on a feeling or principle increases its strength. *If you act on fear, your fear increases; if you act on courage, your courage increases.* As Eleanor Roosevelt said, "You gain strength, courage, and confidence by every experience in which you stop to look fear in the face. You are able to say to yourself, 'I have lived through the horror. I can take the next thing that comes along.'"

Strengthening new assumptions and beliefs is like exercising new muscles. If you are physically out of shape and feeling flabby and lethargic, the way to change your reality is to begin to *act as if you had energy and were in shape* by beginning to walk or jog or bike. When you exercise those muscles, they grow stronger. Similarly, if you are afraid of flying and you *don't* get on an airplane because of that fear, you become more afraid of flying, not less. Acting on the fear makes it stronger.

Most people want to *feel* better before they *do* better. If you ask a depressed person why they are not going out with friends, he may tell you it is because he is too depressed. He is waiting to feel better before he gets involved and has fun. However, the action of going out must be undertaken in spite of depression. You must *do* in order to change how you *feel*.

As you begin to act on your vision, you have a great deal of control over how fast you progress. The more actions you take based on the principles of the person you want to be, the more quickly you become that person.

This might be more clearly understood by looking at the scenario of going to a physical therapist for a bad back. She might tell you, "If you come for therapy twice a week, do exercises three times daily, swim once a week, and buy a new mattress for your bed, your back will probably be better in eight weeks. If you come to therapy once a week, exercise once a day, don't swim, and don't buy a new mattress, it will probably take about six months. If you come to therapy once a week but do nothing else, your back may be better in a year." The speed of recovery in the reinvention process depends largely on the quality and amount of actions taken, just as it does with physical healing.

If you act on fear, doubt, and guilt, your progress will be slow; if you act on confidence, courage, and honesty, your progress will be quicker. You may feel very fragile as you use new behaviors which are in harmony with your reinvented self. In spite of the fragility, you will begin to see great progress towards long-term, positive change.

Facing fears is perhaps the most difficult aspect in any change process. As Charles B. Newcomb said, "There are always two voices sounding in our ears—the voice of fear and the voice of confidence. One is the clamor of the senses, the other is the whisper of the higher self." Fear of the future has the potential to immobilize—keeping people in bad marriages, unfulfilling jobs, self-destructive patterns, and crippling addictions. How can you understand what is happening to you so that you can face your fear fully and move through the transformation process?

The most important thing to remember about fear is that it is not something outside of yourself—it is a behavior you create from within. People often speak of fear as an outside force which *makes* them afraid or stops their progress, but in reality it is a behavior we create. Even the phrase "dealing with the fear" depersonalizes the idea, leading us to assume that it is something "out there" which we have no control over.

Like any other behavior, fear is related to the four basic components: feeling, thought, action, and physiology. As you perceive a situation to be potentially harmful and think fearful thoughts, you also create the feelings and experience the physiology of fear—upset stomach, tight muscles, or shortness of breath. The physiology does not *cause* the fear anymore than sweating causes running. When you add the actions of a fearful person (avoidance, indecision, withdrawal), the behavior is complete and self-perpetuating. Until you change one or more of the components of your total behavior, the fear will continue, growing stronger each time you act on it.

Breaking Up Old Foundations

When you initially act on your vision and face your fears you may feel like you are "going crazy" or even "dying" in a sense. The reinventing process includes the death of those parts of yourself which are not in tune with your intuition and higher self. One of the dominant fears at this point may be that if you dismantle and destroy the old ideas and attitudes on which you have built your life, nothing will be left.

Understanding the urge to dismantle one's self has been a very powerful concept over the years when working with people with suicidal tendencies. Typically, these people have already tried many other behaviors to get their lives to work, but at some point the idea of suicide comes to mind. In the counseling process, I introduce two *different* choices—"killing yourself" and "ending your life," a differentiation which opens the door for reinvention.

For suicidal people, their thoughts and feelings are not necessarily about "ending their lives"—but more about "killing themselves." They have become people whom they dislike intensely, very different from who they want to be or what feels good inside. They want to get rid of these selves and they believe the way to do it is to commit suicide. The urge to "kill oneself" is misunderstood and misinterpreted to mean the same as "ending one's life." The two are quite different. Helping people to see that difference and then make appropriate plans has been very successful. In counseling, I am able to say, "Let's get on with killing this self you hate so you can start feeling good. In fact, you've come to the right place because what I do is help people kill themselves without ending their lives."

At the point that a person is ready to develop a new self, we are ready to define that person and bring him into being through thought and action. For people who are totally focused on the world's power to "make" them feel depressed, finding a way out of the suicidal choice is a very welcome event.

At times, the fear and anxiety which accompany change may feel like some sort of nervous breakdown, but they actually signal a "breaking up." As you move toward a life unified by new principles, ideas, assumptions, and beliefs, the old foundations, like giant ice floes, must break up so that they can be integrated into new structures of thoughts and beliefs.

With this breaking up of old foundations, you may fear that you will lose your ability to function in your personal and/or professional world. This fear is greatest during the transition period when you are no longer the "old" but not yet the "new." To confront this fear and move through it, you must continue the growth process. As Erica Jong says, "I have accepted fear as a part of life—specifically the fear of change. . . I have gone ahead despite the pounding in the heart that says: turn back." Without breaking up what already exists, there can be no new creation.

"Crazy" and fearful thoughts are normal during the transformation period. I assume any self-respecting caterpillar would have similar thoughts as it wraps a grey cocoon around itself in the hopes of becoming a butterfly. Who can begrudge feelings of doubt when you have no guarantee that everything will work out? Your doubt may take the shape of metaphors that mirror your sense of anxiety and confusion as you encounter the unknown. You may dream of falling into an abyss or watching a boat leaving the security of a dock and heading out into an unknown ocean.

Metaphors can become powerful tools further down the road as you decide how you want to resolve these doubts. During the transition period, you may feel helpless and vulnerable, without direction. Part of your growth will be accepting your powerlessness over the dynamics of transformation. Eventually, your ability to gain balance and maintain momentum will depend on taking control of and being responsible for the outcomes you *can* determine. A new understanding of the correct balance between what you can and cannot control in your life is one of the benefits of reinvention.

As a recovering alcoholic, I now know that the changes I have made through my recovery comprise the greatest and most difficult transformation in my life. I was scared, crazy, and depressed during the process and only two things kept me going: remembering how bad things had been while I was drinking and seeing others who had succeeded at what I was now attempting. Having such models is crucial. People in the middle of the reinventing process need to see that it is possible to survive and grow. With these two things in mind, I was able to face my fears and act on my vision—thereby experiencing the freedom, joy, and fulfillment that arise out of the reinventing process.

Chapter 7

_____

Changing Relationships

In many cases, as you go through the transformation
process, those closest to you will be excited that you
are moving closer to meeting your needs and gaining a
sense of self-respect and internal congruence. Though
you may have been living a major part of your life based
on faulty or self-destructive assumptions, many of the
relationships you have formed may still be strong and
functional. As you reinvent yourself you must make some
difficult choices in how to deal with your relationships,

which may be either healthy or dysfunctional. Loved ones with whom you have effective, need-fulfilling relationships will be of tremendous importance as they support and affirm the profound changes you are making. Similarly, dysfunctional relationships may hamper change and deter reinvention. In this chapter I will address some problems and challenges you may encounter in relationships as you proceed through the reinvention process.

Accept that Relationships Change

Many intimate relationships are built on an implied assumption that people in the relationship will remain predictable and similar to the people they were when they were "chosen" as a partner. Any relationship, good or bad, progresses with mutual interdependencies and patterns of giving and getting which have often been worked out with the precision of an interpersonal ballet. If one partner, spouse, or colleague in this delicate balance makes a decision that might be perceived as a threat to the other's well being, the other may choose to become resentful, afraid, or defensive. After all, even when changes are simply the result of repairing or remodeling (like changing the duties expected of each person in a

87

relationship) there are stresses and conflicts. When one partner independently decides to change the basic principles and assumptions by which he or she lives, the stress in the relationship soars.

Because those close to you may be threatened by your changing, you may perceive that many of your important and meaningful relationships are jeopardized. It may appear that the more you attempt to resolve your pain, the more resentful others become. While you may perceive yourself as simply wanting to change what you "can't live with anymore," others may seem to be determined to hold you in your old roles and expectations. Both you and your loved ones need to understand that regardless of whether you had chosen to reinvent yourself or to continue on a self-destructive path, your relationships would not have remained the same. The dynamics of any relationship are constantly changing. Your reinvention provides the opportunity to create a new and positive direction in your relationships with others if they are willing to accept that change.

Communicate! Communicate! Communicate!

Choosing to take the reinvention path—often mirrored in others' frustrations as "I don't know what's wrong with you these days"—can be very threatening. Friends and loved ones may become anxious because they seem to have no control over what's going on. They may perceive you as "running the show" or "doing this *to* them." What you may overlook is that, as you resolve your life in new and different ways, they may feel their security is falling apart. The best you can do in this situation is communicate your perceptions of what is going on in your life and why you believe the reinventing process is the only healthy course to take.

Commit to Shared Values, Not Specific Results

In situations where others are inexperienced regarding transformation, they will often be most concerned about "how things will turn out—what the *results* of your behavior will be." Any description of the future that you offer will be inadequate because this central issue (what *form* the future will take) cannot be specifically or accurately answered. In acting as your reinvented self, you are putting your energy into controlling who you are and what you do so that you are aligned with your inner truth (intuition) and your higher self (the person you want to be). You are *not* focusing on the results of your behaviors and how they will affect people, places, and things. If you measure your behavior by what feels intuitively right, by what you hear from your inner voice, and by how you adjust your behavior to follow internal instructions, you cannot predict the exact outcomes of your transformation. *Who you are* in the world will be more important than *what is happening* in the world.

Not being able to guarantee the future does not mean that you can't plan or make commitments. On the contrary, you may have a clearer *direction* than you have ever had. You can provide loved ones with a general compass direction, but not road-map specificity. You can make promises and commitments based on "good faith," but you cannot honestly guarantee specifics. If you accept the process of living in harmony with your intuition, you will also let go of the fantasy that you can guarantee the future.

The most important ingredient in restructuring a relationship after one participant has decided to reinvent himself is a mutual commitment to living in tune with intuitive signals and basic needs. When both people actualize this commitment as a primary goal in the relationship, love and energy is renewed. When you, as an individual, become stronger, the relationship benefits from that growth.

As you proceed with your reinvention, explore with your loved ones the direction in which you are taking your life and where they fit into the picture. Positive and profound change is very exciting—those you care about can benefit from this energy if you both commit to shared values and beliefs.

Resolve the Issue of Selfishness

As others feel the rules of the relationship changing, they may begin to see themselves as oppressed, unsuspecting victims of selfish, spontaneous, and non-negotiated behavior. They may assume the role of victim, leaving only one other role open in the relationship—the role of victimizer. When placed in this situation, it is possible to convince yourself that you should feel guilty for your actions. If you accept this role, built on the paradigm that you are responsible for other people's happiness, you may feel so guilty that you rethink your doubts about transformation.

Unfortunately, by the time many of us have come to the point of changing self-destructive and dysfunctional patterns in our lives, some of the relationships we are in are also dysfunctional. If this is true, dealing with our own unhealthy patterns may send messages to people with whom we are involved that they must now deal with theirs. No one wants to change on someone else's timetable, and very often people choosing transformation will see many manipulative behaviors from loved ones who are attempting to maintain the status quo.

One powerful controlling behavior used by people who feel threatened by transformation is to hurt *themselves* in some way so that we will be "forced" to tend to them rather than to ourselves. These self-destructive behaviors are very real and can be very damaging. They may include becoming sick, abusing drugs or alcohol, having a nervous breakdown, creating an intense financial crisis, focusing on or creating emotional crises with other friends or family, or even contemplating suicide. All of these are attempts to gain control by convincing us that we should not continue on our uncertain path. These behaviors are not necessarily chosen consciously, and insinuating that they are will probably be received with outrage.

As loved ones choose self-destructive behaviors, you may want to stop your progress and help them. However, in the throes of reinventing yourself, it is crucial to continue your progress and *face your fears*. If you don't, you can easily slip back into old and destructive behaviors which are more familiar. To ensure your own survival and development, you must allow others to experience the consequences of their self-destructive behaviors. Remember that they have as many resources for help as you do and that they can avail themselves of that help at any time. The message you may be getting from them is that *you* are the only one who can help. This can be difficult, especially if you believe them. This dysfunctional belief may still be strong enough to seduce you into slowing your progress.

You must remember that there is nothing wrong with helping loved ones, but you must make an effort to do it within the context of continuing the reinventing process. *You must understand that your desire to reinvent yourself is an unselfish act.* The desire to grow and live according to your inner voice is a natural instinct. As Amrit Desai, founder of the Kripalu Center for Yoga and Health states:

> Everything you do, all the actions you perform—if they lead you to your inner source (even though they may appear selfish)—they are selfless actions. Because that makes you free from dependence on others; for the dependence on others becomes the source of all selfishness. Dependence on others means that now you have lost your center and you are searching for it everywhere except within yourself.

In reinventing yourself you attempt to act more in concert with your intuitive signals. If you follow the dictates of your intuition, you will eventually come to a point where you are more loving, more powerful, more playful, and more free. The same process is open to your loved ones.

Find Mutual, as Well as Individual, Support

You may also need help and support from people outside of the relationship. Marriage or relationship counseling can be beneficial, especially if the counselor is one who understands the dynamics of transformation and is willing to focus on the function and process of the relationship rather than its eventual form. If the counselor leads you only to discussions about past problems or a specific future, he or she will probably not be of much help to you on your intuition-directed journey. In the counseling process, it is important to talk about the present and the direction in which you want to take your relationship.

To maintain your focus and strength during these difficult transitions and at the same time stay positively and lovingly involved with others, you must walk a very fine line. Listen to the concerns of your friends and family, sharing with them the content and process of your journey. Your major energies must still center on your own reinvention, but this does not preclude doing what you can to give support to friends and loved ones. You must evolve through your own confusion before you can "be there" for others, but if you provide those around you with information as you gain it, they may be better able to make choices which allow them to keep their own balance. To do this you will need to develop new attitudes and behaviors. Learn to detach from their pain without withdrawing your love and concern for them. Communicate this love and concern and share with them your hopes for the future. Ask that they help you by having faith—and help themselves by taking care of their own needs. You must *choose* to believe that they are responsible for how they deal with the signals they are experiencing, just as you are trying to be responsible for your own. Attempt to "let go" of the outcomes of your relationship and focus on the quality of your interaction, maintaining a clear understanding of the distinction between "letting go" and "giving up." Reinventing yourself will strengthen a healthy relationship, but changing a relationship will always take strength.

Chapter 8

Tools for Reinvention

I have always marveled at the simplicity of truth. When I hear something about life that resonates with authenticity, I often think, "That's just common sense." My mind finally understands something that I already knew in my heart. But what I know to be true is not always what I act on. I forget what I know when I need most to remember. The idea of leading a life where *everything* I do is healthy seems overwhelming. This does not mean that I am not willing to move toward being more honest about what I know and what I do. As Alcoholics Anonymous tells us:

> No one among us has been able to maintain anything like perfect adherence to these principles (twelve steps). We are not saints. The point is we are willing to grow along spiritual lines. The principles we have set down are guides to progress. We claim spiritual progress rather than spiritual perfection.

The same is true as you proceed through reinventing yourself. Progress is all you can expect. Focusing on perfection is unrealistic and self-defeating. A realistic expectation is that you will move toward more clarity, strength, and happiness.

As you move into an unknown future, you need guidance to help choose behaviors other than fear, disappointment, doubt, anger, and impatience. This guidance comes from others who have gone through a similar process as well as from your own changing attitudes and awareness. Your actions are the bricks with which you finally build your life, but your attitudes and beliefs are the tools which you use to put these bricks together. Six such tools follow.

1. We all need a little help from our friends.

In the midst of transformation it is important to receive support from people who have done what you fear doing. If you can connect with others who have made similar changes in their lives, you will be better able to create an acceptable future. This is true for alcoholics, adults considering adoption, cancer patients, couples ending marriages, couples considering marriages—anyone who is stepping into the unknown and looking for ways to measure the wisdom of the choices before them.

I have many "ex-religious" friends (nuns, brothers, priests) whose decisions to leave their respective religious communities were the most difficult and transforming experiences in their lives. In making the change, they were forced to question every assumption they had about who they were and how they could still lead meaningful and spiritual lives. They struggled with the very essence of what it meant for them to have committed themselves to a life with God and what they should maintain or reject as they left their religious orders. How could they now call themselves honest? Who were they now and who could they become? Could they keep their promises to the church, to God, and to themselves if they reassessed their lives and made a different choice? My friends all made the transition, but not without tremendous fear, struggle, and pain. Many of them encountered at least one other person who had "survived" the decision with which they were struggling. They met people who shared their own anguish, who had walked into their unknown future and were better for it.

To successfully reinvent yourself, you need information and support from people who have accomplished major transformations in their lives. Without these models, you will feel even more alone in your struggle. Comparing your reality with others who have *not* gone through this process will create even more fear. As you look for mentors, remember Abraham Maslow's words, "You cannot perceive what you are not. It cannot communicate itself to you." Seek guidance from people who have faced their fears. They know what a successful journey is because they have made one themselves. They have made essential changes in beliefs, values, assumptions, and behaviors and found self-fulfillment. They have experienced the transformational process and can help you understand its ramifications.

As you look for people to support you, fight the fearful thought that asking for help is selfish and an intrusion into others' lives. Mentors, guides, or sponsors who have successfully reinvented their own lives will not see your request for help as an imposition—their assistance will be an extension of their own growth and they will welcome the opportunity. Find people who are where you want to be and ask them how to get there. If you choose the right people, they will not tell you what you should do; they will help you ask the right questions in order to get to a "place" that is right for you. If they are angry and judgmental toward you for asking, you were probably wrong about wanting to be where they are.

2. Focus on your vision and validate yourself for what you are attempting.

"When you're up to your ass in alligators," the old saying goes, "it's hard to remember that your initial objective was to drain the swamp." In the process of profound change, remember that you are changing your life because of a desire to make things better, not worse.

The immense challenge of changing basic foundational thinking can be easily underestimated. First you may struggle to hear your intuition. Then you may struggle to figure out what to do about those signals and wonder whether you have the courage to act. That's a lot of struggling.

Changing what you do can be difficult, changing *why* you do it can be even more difficult. During times of doubt, remember that nothing of value is achieved without great effort. By the time you come to your moment of truth you may have invested massive amounts of time, energy, thought, and resources into beliefs you are now discarding. Many of us have gone to great lengths to avoid the journey we are now taking, and we may justifiably feel that we are risking what we have without a better outcome assured. This is truly courageous and you must learn to validate yourself for these choices.

Yes, you are courageous and . . .

. . . so was I when I was dealing with my divorce, meeting with the lawyers and preparing myself to be who I wanted to be without making that contingent on my ex-wife's behavior. I had to face old assumptions about being strong and loving—assumptions that had their roots in obligation, caretaking, and the desire to avoid conflict.

. . . so is the wife who decides she must risk losing the security of a marriage in which she is constantly pressured to relinquish her self-respect. She will grapple with the issues of commitment, her role as a woman, and the boundaries of her duty to provide a stable home for her children.

. . . so is the son who stands up to his abusive father because he knows he won't be able to live with himself if his little sister gets hurt. He may challenge his entire system of values around family loyalty, physical fear, and parental love.

. . . so is the teacher who teaches to the energy of her students rather than what the curriculum mandates. She will struggle with her expectations of what she is *supposed* to teach, what it means to be honest about her choice, and whether her professional responsibility is to listen to her heart or her principal.

. . . so is the alcoholic husband who goes to endless meetings in his first months of sobriety while his family complains that he is never around at all—that he was "more of a dad when he was drinking." He must face his fears of acting on his faith and putting himself "before" his family.

. . . so is the woman who admits that she has a drug problem and goes to a treatment center. All of her life she may have believed that she could handle her problems alone, that asking for help showed weakness, and that willpower was enough to do anything. By asking for help she questions the foundations of who she is.

We are all courageous in any situation, no matter how small, in which we challenge ourselves to live according to the dictates of our inner voices rather than our fear.

3. When it seems too hard, ask yourself "Compared to what?"

There will be moments in the journey when you will say to yourself, "I can't do this," or "I didn't realize it would be this difficult." At these times remember that you are making a choice based on *real* alternatives in the *real* world. *Going forward may be terrifying, but it is slightly less terrifying than going back or standing still.* Recall your moment of truth when you finally admitted that your previous path was ineffective and that you were headed for certain destruction. True, the path of change is risky and the old path was certain—but it was also unworkable. *We are not changing because we want to, we are changing because the old way wasn't working.*

Compulsive overeaters, for example, must remember that by the time they are willing to risk abstinence they have exhausted all other avenues for avoiding the hassles of eating. They have changed *what* they ate, *where* they ate, *who* they ate with, and *how* and *when* they ate. Manipulating people, places, and things as a way to be happy is ineffective—as is living without regard for their basic needs. Abstinent eating is usually not a choice taken willingly or gladly; it is the only alternative left other than self-destruction through compulsive overeating.

When you decide to reinvent rather than repair or remodel, it is only because you have already searched desperately for a choice which has few or no negative consequences. Unfortunately, such a solution does not exist.

Finding our balance during difficult times involves accepting that every solution has its problems. If we ask for a raise, our boss will be upset; if we don't, our spouse will be upset. If we move to Arizona we will have to move away from one set of grandparents; if we don't move we will be choosing to live away from the others. If we teach kids what they want to know, they may not learn what they are supposed to learn; if we teach them what they are supposed to know, they may not choose to learn it. *We don't get to choose whether or not we have problems; we only get to choose which problems we want to live with.*

Learn to "look back, but don't stare." There is no quicker way to become depressed than to compare your progress to where you *should be* rather than where you have been. In looking forward you may focus on how far you have to go and how slow you are going. Looking back allows you to focus on your growth. You can feel grateful for the opportunity now before you. Be careful not to look back too far, however, or you may only remember the initial good times and how well the ineffective behaviors worked *before they stopped working*. If you are leaving an intimate relationship that has soured, it is important to remember how bad things had become in the last months of the relationship. Otherwise, you may become depressed and lose the momentum needed to move forward. Accept that there is no acceptable past to go back to and remind yourself of the progress you have made.

4. Long-term changes begin with small steps.

Making successful long-term changes is a process which proceeds relatively slowly and from a solid base of understanding. When you take your first steps in a new direction, you are not as balanced as you will come to be. Take small solid steps when starting the process. If you stretch too far, you can easily lose your balance, slip, and fall. A long hill with treacherous footing is most surely negotiated by steps taken with careful planning and pacing. The same is true of acting on your vision in the reinventing process.

In the throes of the change process, you will feel anxious. You will want to resolve that anxiety as soon as possible, and the impulse will be to move quickly through the process—to take bigger steps. If you try to do too much too fast, you will slip and lose more ground than if you had just kept going one small step at a time. This will allow you to experience success with new values, beliefs, and behaviors. Confidence allows you to continue taking risks and believing in your new self. Perceiving yourself as a successful person is also one of the outcomes of transformation, and as each small step is taken, you become more confident in your ability to move forward.

Every experience, whether it be large or small, provides an opportunity to use your new principles, values, and behaviors. Whether the *incident* is how much to eat for dinner, whether to take a new job, or how to confront a loved one about a problem, the *issue* will always be whether you choose to behave as your old self or as your reinvented self. Transformation is not accomplished through big glamorous changes in rapid succession. In fact, long-term change is the result of the little choices, the tiny steps that you take when no one else is looking.

Carol, a young woman, wanted to reinvent herself because she had decided that she was no longer willing to be treated as "the baby in the family." She was ready to start making choices unified around the values of independence, assertiveness, and fun. She hoped to do this quickly, but didn't factor in that changing the patterns of nineteen years takes time, skill, and patience.

Carol wanted quick progress so she took huge steps. She decided that an independent, assertive, fun-loving Carol would immediately move out of her parents' home and get her own apartment with a group of friends. Rather than seeing that as a long-term goal, she left home two days later and moved in with a group of older women she knew from work. She told her dad that she was "grown up" now and that he should stop telling her what to do. She changed her behavior before she had a good support system for the "new Carol," and within a month she realized that she had moved too fast. The women she lived with were all busy with a social crowd that went to nightclubs where she was too young to go. She also found that she no longer had the daily support and love from her family that she had taken for granted. She felt incredibly lonely.

Before changing everything, Carol should have experimented with her reinvented self until she gained some proficiency as the "new Carol in the old situation." She might have asked, "How would an assertive Carol handle her father when he told her she needed to be in by twelve o'clock? How would an assertive, fun-loving Carol handle her boyfriend when he wanted to sit home and watch TV all night? How would an independent Carol handle Christmas?" She might have changed her behavior in little steps, like having more fun at her job and in her relationship with her brothers and her boyfriend.

Carol took too big a step and lost her balance. She became depressed and felt sorry for herself. Two months later Carol was back in her parents' house, feeling beaten and demoralized. All her grand plans had fallen through. "I really am the baby of the family," she said to herself, "I guess I'll never be independent." If Carol had gone slowly, using her reinvented behaviors in simple situations before she took on more complicated ones, she would have set a much stronger foundation for her new self.

As you choose how far or fast you progress in your life, it is important to treat foundational change as a long-term goal rather than something to accomplish immediately. Imagine a father putting his young children to bed one night and one of them saying, "I only want Mommy to put me to bed." Hearing this, he tells himself that he has been an uncommitted parent who has been selfishly pursuing his career while staying "at arms' length" from his kids' daily lives. He decides that the person he wants to be is loving, unselfish, and committed. At this point, he should refrain from immediately trying to do *everything* that person would do; instead he needs to focus on the next step. In the diagram below, he must remember that he starts at Point A, that his ultimate goal is Point Z, but his next step is Point B.

Starting point

Using the situation above, Point A represents the father's decision to be more *loving, unselfish,* and *committed.* His idealized vision (Point Z) is that if he were *really* that person, he would be home early every night to help with the kids' dinner, spend weekends with his kids (rather than working), become a little-league coach, and leave his work at the office. If he tries to do all these things at once he may antagonize everyone unnecessarily, including his kids who already have a daily routine. Instead of becoming more committed, he may end up adding *overbearing* and *impatient* to his list of undesired personal qualities. Moving in the direction of his ideal self must happen slowly—there are many realities that need to be adjusted. His wife needs to adjust to his sharing in decisions about the kids' daily schedules, his boss has to adjust to an employee who is no longer willing to commit all of his time to work, and he has to adjust to a different pace of life and fulfilling his work commitments in new ways.

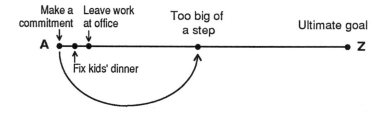

His best choice at this point would be to take the next small step forward. This might mean coming home early from work *one night* during the next week. It might also mean finding a few hours during the weekend when he could spend time with the kids, and he might even go so far as to leave his briefcase at work one night. If he makes too many changes too quickly, he may lose his balance in the process and feel overwhelmed by all he has to adjust to at once. However, taking one small step at a time will help him gradually become the parent he wants to be.

We often believe that we have to move quickly because there is so much time to make up. Upon being overlooked for a deserved promotion, a previously successful employee now realizes that she has wasted three years being intimidated by her supervisor. A young man whose early years included many wonderful personal relationships now sees the need to end a destructive relationship and feels that he has thrown away the last five years of his life. A recovering alcoholic, who once had a healthy social life, sees the wreckage of eight years of alcoholic drinking and realizes that he now has no real social life at all. In a situation similar to these, you may tell yourself that you have been moving in an ineffective direction for so long that it will surely take a long time to get to where you want to be. The image in your mind is that at some point you were headed in a healthy direction (at Point A) and then spent three, five, or eight years messing things up and moving off course (to Point B). You may worry that you are now a long way from where you want to be, and that it will take another three, five, or eight years to get back on the right track (to Point C).

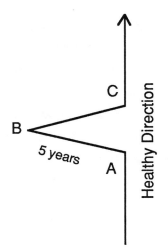

The reality is that you probably moved from Point A, to Point B, and then spent the rest of the time going around in circles.

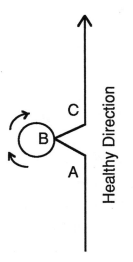

Yes, you must stop your whirling about and get yourself back to a point where your life is working, but the distance is not nearly as far as you imagine. You do need to find a way to make healthy changes in how you think and act, but once you stop repeating the old destructive patterns (going around in circles), you may find that in a few months you can regain the control you once had. By taking small, solid steps, the road back is often shorter than you imagine.

5. Develop positive metaphors.

Although it seems paradoxical, you can sometimes use your mind to help you go *beyond* your own thinking. Your mind is where you talk to yourself—where you use language and metaphors to help you understand the world. You can use your mind to help create sense in those moments when you feel you have none, especially when you begin to act on your vision and face your fear.

In the confusing times that accompany a profound shift in life's values, we are often left without a clear sense of direction. One way to find that direction is by creating a metaphor for yourself—an allegory in which you can participate. In times where you were confused and upset you may have inadvertently compared yourself

to a tiny ship being tossed around in a hurricane, a balle-
rina loaded down with sandbags, a mountain climber who
never reaches the highest peaks, or someone falling
through space. Through these metaphors, you reflect
your thoughts and feelings in current situations. They
reflect feelings of being out of control, lost, or overwhelmed. In order to create strength in the throes of the
reinventing process, you can also develop metaphors
which will help focus on your positive qualities and provide support.

Metaphors of weakness do no good when choosing a
direction, but metaphors of strength do. If you see yourself as a leaf being buffeted around by the wind, the
metaphor is one of powerlessness. To ask yourself,
"What would the leaf do at this point?" would not be
helpful. The leaf can do nothing but wait for the wind to
blow and hope for the best. This metaphor only perpetu-
ates feelings of powerlessness. However, you can gain
strength by describing your struggle in terms of a positive, proactive metaphor. For example, if you describe
yourself as an eagle soaring over a vast terrain, the
answer to "What would the eagle do at this point?" gives
some direction. You can visualize the eagle being in
control, detached, and confident. The eagle gives a vision
of transcendence, the leaf a vision of passivity. As you
reinvent yourself, your metaphors will help shape your
journey.

Another powerful metaphor is a caterpillar turning into a butterfly. All the components are there for giving direction in moments of doubt: the outcome is beautiful and life-fulfilling, there is a time of doubt which demands waiting and courage, and there are opportunities for action which include climbing up a branch, spinning the chrysalis (which eventually becomes both a tomb and a womb), and breaking out of the chrysalis when the time is right. Some of us find we can return to religious or mythical metaphors that we used as children, allowing us to view the process as positive. We may choose the metaphor of Job showing allegiance to God's will or the mythical Phoenix rising from ashes. These metaphors help us see the validity of our journey and may even clarify our feelings and thoughts to others.

The metaphor you choose is up to you. If you want to encourage and nourish yourself, you need a mental picture of this journey which reflects courage, positive transition, and overcoming doubt and fatigue. Focus on the importance of the positive transition process, rather than the outcome, and on eventual victory over fear and adversity.

6. Proceed with faith; the path will open up as you move forward.

One of the biggest obstacles to successful change is the desire to be certain of the results. We want to know the outcomes of all of our actions so that we can maintain control and not be caught off guard. Evaluating your choices according to intuition demands that you take things one step at a time. The most effective strategy to use in confronting your fears is simply to complete whatever task lies in front of you. This is hard, because you want to look further down the path, see what comes four steps from now, and know how things will work out.

As children we learn to create specific ideas of what we want. We also learn to fear situations where we might not get what we want. We believe incorrectly that what we *want* is the same as what we *need*. It is only as adults that we learn we can get what we need even if we don't get what we specifically want. Regardless, we often carry this fear into adulthood, which may slow our ability to change.

Peter's story illustrates the importance of this point. Peter was debating with himself about whether to run for the local school board and was being encouraged by friends to do so. He had been experiencing a sense of civic duty and his instincts told him that he didn't feel good complaining about the school system without also taking some responsibility. Peter wanted to be someone who took an active role in the community, but there were so many unknowns. Would he be able to understand all the federal regulations? Would it take him away from his family too often? Would his kids feel awkward because he was on the school board? Was he in touch enough with the community to be on the board? Would there be people on the board who intimidated him? What would he do about a second term? . . . By the time he finished stirring up all of his fears, it hardly seemed worthwhile to even consider the prospect.

However, Peter did want to be more civic-minded and did not want his decisions unified by fear. Peter was immobilizing himself because of future uncertainties. His mind wanted guarantees about the future and his intuition only told him about the present.

When struggling with learning how to listen to your intuition, and when unsure of what message to follow, it is important to do two things. First, *stop* trying to answer questions so large that they immobilize you. Second, *start* focusing on the step that is directly in front of you. In Peter's case, he stopped trying to make the decision "to run or not to run"—a choice he could not make at this time—and he *started* making many smaller decisions that he could make. Peter could move through his fear by asking himself, "What am I clear about doing? What parts of this decision am I *not* conflicted about? What

does my intuition tell me is the right thing to do *next*?"
This may be something as small as asking his friends why
they think he would do a good job. It may mean attending
one school board meeting and seeing what goes on there,
or it may mean talking to his daughter about her thoughts
on the subject. *Any step you take based on the values you
would like to strengthen in your life will help you
develop the courage to take future steps.* You must do
what you are clear about and let go of what you are not.

As you allow patience and trust to be the unifying
principles of your behavior, you give yourself time to
grow. Your intuition-directed journey will become clear
one step at a time, and you can be sure that if you take
the step in front of you, the next one will be clear when it
needs to be.

Peter eventually learned to focus only on his next
step, and he even created a metaphor to help. He saw
himself as a traveler in outer space walking through a
large transparent tube which had no bottom. As he
moved forward through this tube, a solid step would
appear beneath his foot in the moment just before he
imagined falling through the tube and into endless space.
The step he moved from and the step he moved to were
the only ones there, for each time he moved forward, the
step he had just moved off would disappear and become
the one in front of him. He couldn't go back, and in
looking forward he didn't know whether there would be

a step ahead of him when he needed it. What he had to do, though, was to take his foot off the previous step so that it could become his next. He had to act on faith that it would be there. This metaphor helped Peter feel that he had more control. He had created the metaphor of a traveler who acted on the *faith* that he would always have a next step if he had the *courage* to leave his last. Therefore, he felt good about his willingness to proceed on faith rather than fear.

These six tools will make your journey easier because each is solution-oriented and an actualization of faith. Faith will always be a necessary component of the reinventing process, because having faith is the most positive thing you can do when you are unsure of the future. The next step towards a need-fulfilling life will become clear when you are ready. This has been validated by the experience of millions of people—people in twelve-step recovery programs, people in love, people dying or having near death experiences, people in spiritual communities and religious organizations, people surviving all forms of personal crises.

Reinventing yourself is an intuition-directed transformation which allows you to discover and experience the love, power, fun, and freedom within you. This process provides you with ultimate freedom—the ability to be happy in the world without needing to control or be controlled by other people, places, or things. The reinvention path—choosing a way of living based on the values and beliefs of your higher self—is not easy, but it does lead you on the exciting journey of becoming the person you want to be.

Epilogue

Writing this book has allowed me to challenge myself in ways that I was not willing to try before. This book is the product of another level of my reinvention in which I created the thoughts and actions of the "person I wanted to be." This reinvented person is someone who wants to feel *competent* as a writer, *courageous* enough to stick with the project, and *faithful* to the process of creativity that I know is difficult. Most of the thoughts I invented to accompany these values sounded like variations of "I think I can, I think I can..." and actions included making the time to write, asking for feedback from friends and colleagues, and continuing to slice away hundreds of words of writing that seemed so interesting days and months before.

As I finish these chapters, I am reminded of my colleague's words, "Don't tell people it's so hard; they won't want to do it." I understand what she is saying, but I also believe that the results of moving toward harmony with our higher selves will far outweigh any pain we go through. The joy of a clear path through the myriad of important decisions that must be made in developing a quality life, the sense of pride and courage that comes from facing fears that have stopped us before, the excitement of learning new behaviors that allow us to be free in any situation we encounter are only part of the blessings we enjoy as we learn to listen to our intuition. As Nikos Kazantzakis said, "As I watched the sea gull, I thought, 'That's the road to take, find the absolute rhythm and follow it with absolute trust.'"

Thank you for reading this book and for the time you have taken to share this world with me. There is an important truth that I try to remember as I lead my life. It helps me to maintain my clarity and to focus on what *is* significant rather than what *seems* significant. I want to leave you with this thought and the words that come to mind when I apply it to myself. I hope it helps you become more fully "the person you want to be."

WE ARE NOT HUMAN BEINGS GOING THROUGH A SPIRITUAL EXPERIENCE; WE ARE SPIRITUAL BEINGS GOING THROUGH A HUMAN EXPERIENCE.

When I doubt that what I know
intuitively makes sense,
I return to these words.

When I try to define every part of the process of
living so that I can finally and completely say,
"I understand it all,"
I return to these words.

When I forget that I am not supposed to be perfect,
I return to these words.

When I fear that talking about intuition and
psychological/spiritual development may
upset or offend some people who would
rather see the world another way,
I return to these words.

When I forget that the process of transformation is a
gift to enjoy rather than a hardship to be endured,
I return to these words.

And when I feel hurt and scared,
and begin to believe that I am alone
without guidance in the universe,
I return to these words,
and I am grateful.

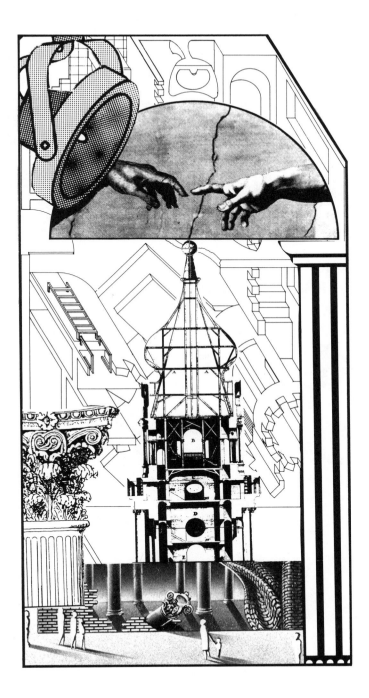

Bibliography

Alcoholics Anonymous. New York: AA World Services, 1976.

Beattie, Melodie. *Co-Dependents' Guide to the Twelve Steps*. New York: Prentice-Hall, 1990.

Covey, Stephen M. *Principle-Centered Leadership*. New York: Simon & Schuster, 1990.

———. *Seven Habits of Highly Effective People*. New York: Simon & Schuster, 1989.

Dean, Amy. *Night Light*. Center City, MN: Hazelden Foundation, 1986.

Desai, Amrit. *Working Miracles of Love*. Lenox, MA: Kripalu Publications, 1985.

Ferguson, Marilyn. *The Aquarian Conspiracy*. Los Angeles: Jeremy Tarcher, 1987.

For Today. Torrence, CA: Overeaters Anonymous, Inc., 1982.

Ford, Ed. *Freedom From Stress*. Scottsdale, AZ: Brandt Publishing Co., 1989.

Gawain, Shakti. *Living In the Light*. Mill Valley, CA: Whatever Publishing, 1986.

Glasser, Naomi, Ed. *Control Theory in the Practice of Reality Therapy*. New York: HarperCollins Publishers, 1989.

Glasser, William, M.D. *Control Theory*. New York: HarperCollins Publishers, 1984.

————. *Positive Addiction*. New York: HarperCollins Publishers, 1976.

————. *The Quality School*. New York: HarperCollins Publishers, 1992.

————. *Reality Therapy. A New Approach to Psychiatry*. New York: HarperCollins Publishers, 1965.

————. *Stations of the Mind: New Directions for Reality Therapy*. New York: HarperCollins Publishers, 1981.

Good, E. Perry. *In Pursuit of Happiness*. Chapel Hill, NC: New View Publications, 1986.

Hay, Louise. *You Can Heal Your Life*. Santa Monica: Hay House, 1984.

James, Jennifer, Ph.D. *Success Is the Quality of Your Journey*. New York: New Market Press, 1983.

Kaufman, Barry Neil. *Giant Steps*. New York: Fawcett Press, 1979.

McWilliams, Peter and Roger, John. *You Can't Afford the Luxury of a Negative Thought*. Prelude Press, 1988.

Overeaters Anonymous. Torrance, CA: Overeaters Anonymous, 1980.

Peck, M. Scott. *The Road Less Traveled*. New York: Simon & Schuster, 1978.

Powers, William. *Behavior: The Control of Perception*. Chicago: Aldine Press, 1973.

Satir, Virginia. *Self-Esteem*. Milbrae, CA: Celestial Arts, 1970.

Siegel, Bernie S., M.D. *Love, Medicine, and Miracles.* New York: HarperCollins Publishers, 1986.

Silverstein, Lee M. *Consider the Alternative.* Minneapolis: CompCare Publications, 1977.

Touchstones. Center City, MN: Hazelden Foundation, 1986

Vaughn, Frances and Walsh, Roger, Eds. *Accept This Gift.* New York: St. Martin's Press, 1987.

Zambucka, Kristin. *Ano Ano the Seed.* Honolulu: Mana Publishing Co., 1984.

About the Author

Dr. Barnes Boffey's professional life reveals a diversity of work and interests. After receiving a B.A. in Drama from Middlebury College, he went on to teach elementary school, and from there to receive an Ed.D. from the University of Massachusetts. He continued a career in education which took him to the University of Cincinnati and eventually to Dartmouth College where he served as the Director of Teacher Preparation. His experiences also include twenty years of parenting, directing a boy's summer camp, and maintaining a private practice for individual, family, and marriage counseling.

Dr. Boffey is a senior faculty member of the Institute for Reality Therapy. He conducts workshops and seminars in Reality Therapy and Control Theory for residential youth programs, drug and alcohol programs, parent groups, and public and private schools throughout the United States as well as training in Canada, Slovenia, Croatia, and Australia. He also leads intensive training weeks for professionals interested in the application of Reality Therapy and Control Theory.